Introducing Game Theory and Its Applications

DISCRETE MATHEMATICS AND ITS APPLICATIONS

Series Editor

Kenneth H. Rosen, Ph.D.

AT&T Laboratories
Middletown, New Jersey

Richard A. Mollin, Algebraic Number Theory

Richard A. Mollin, Fundamental Number Theory with Applications

Richard A. Mollin, An Introduction to Cryptography

Richard A. Mollin, Quadratics

Richard A. Mollin, RSA and Public-Key Cryptography

Kenneth H. Rosen, Handbook of Discrete and Combinatorial Mathematics

Douglas R. Shier and K.T. Wallenius, Applied Mathematical Modeling: A Multidisciplinary Approach

Douglas R. Stinson, Cryptography: Theory and Practice, Second Edition

Roberto Togneri and Christopher J. deSilva, Fundamentals of Information Theory and Coding Design

Lawrence C. Washington, Elliptic Curves: Number Theory and Cryptography

DISCRETE MATHEMATICS AND ITS APPLICATIONS
Series Editor KENNETH H. ROSEN

Introducing Game Theory and Its Applications

Elliott Mendelson

CHAPMAN & HALL/CRC

A CRC Press Company
Boca Raton   London   New York   Washington, D.C.

Library of Congress Cataloging-in-Publication Data

Mendelson, Elliott
    Introducing game theory and its applications/Elliott Mendelson.
          p.   cm. — (Discrete mathematics and its applications; 28)
    Includes bibliographical references and index.
    ISBN 1-58488-300-6 (alk. paper)
    1. Game theory. I. Title. II. CRC press series on discrete mathematics and its applications; 28.
QA269.M455 2004
519.3—dc22   2004042807

Visit the CRC Press Web site at www.crcpress.com

© 2004 by CRC Press LLC

No claim to original U.S. Government works
International Standard Book Number 1-58488-300-6
Library of Congress Card Number 2004042807
Printed in the United States of America   2 3 4 5 6 7 8 9 0
Printed on acid-free paper

Preface

This book is intended to be an easy-to-read introduction to the basic ideas and techniques of game theory. It can be used as a class textbook or for self-study. It also could be helpful for students who must learn some game theory in a course in a related subject (such as microeconomics) and have limited mathematical background.

In recent years, and especially since the Nobel Prize in Economics was awarded in 1994 to John C. Harsanyi, John F. Nash Jr., and Reinhard Selten for their research in game theory, people have been intrigued by the unusual connection between games and mathematics, and the author hopes that this curiosity may be satisfied to some extent by reading this book. The book also will prepare them for deeper study of applications of game theory in economics and business and in the physical, biological, and social sciences.

The first part of the text (Chapter 1) is devoted to combinatorial games. These games tend to be more recreational in nature and include board games like chess and checkers (and even a simple children's game like Tic-Tac-Toe). There are also many such games that are challenging even for accomplished mathematicians and our study covers various clever techniques for successful play. The rest of the book deals with the general theory of games. Chapters 2 and 3 contain a thorough treatment of two-person zero-sum games and their solution, which is the most well-understood part of game theory and was developed from the 1920s through the 1950s. John von Neumann was responsible, almost by himself, for inventing the subject, which reached its climax with the elaboration of the simplex method. Chapter 4 introduces the reader to games that are not zero-sum and/or involve more than two players. Here it is often natural to consider cases where the players must no longer act as isolated individuals but are permitted to form coalitions with other players. Games in which this happens are called cooperative games. All of this is an area of current research and there is still no general consensus about its concepts, methods, and applications. Here we have only attempted to present the basic ideas and some of the less controversial results, so that readers can venture further on their own. The reader will find in this chapter some of the applications that make game theory so interesting, for example, in economics, in the theory of political power, and in evolutionary biology. We assume no previous acquaintance with the details of these subjects. Our treatment is intended to provide enough of the fundamental concepts and techniques of game theory to make it easier to understand more advanced applications.

There are three Appendices. Appendix 1 reviews finite probability theory and provides whatever is necessary to understand all applications of probability in the book. Appendix 2 sketches an axiomatic treatment of utility theory. Although the utility concept is necessary for a proper understanding of game theory (and economics in general), we have avoided dealing with the subject in the body of the text because it would be a difficult and unnecessary distraction for the readers in their first confrontation with game theory. Appendix 3 contains two proofs of Nash's Theorem on the existence of Nash equilibria. The reason for consigning them to an appendix is that they depend on sophisticated results from topology, with which many readers will not be familiar.

The section of Answers to Selected Exercises contains brief solutions to enough of the exercises for the readers to be sure that they understand what is going on, and then usually leaves some exercises to be answered on their own. There is an extensive Bibliography that contains not only the articles and books referred to in the text, but also many readings that may attract the readers' interest and extend their knowledge.

The author owes a great debt to Professor Alan Sultan for advice on the theory of linear programming and for the masterful exposition in his book, *Linear Programming* (Academic Press, 1993). He also is grateful for the help given by Bob Stern and the other editors at CRC Press. Thanks are due, above all, to the author's wife, Arlene, for her patience and understanding during the trying period of gestation and writing that every author's family endures.

Roslyn, New York
March 2, 2004

Contents

Introduction

Game theory is a mathematical study of games. By a *game* we mean not only recreational games like chess or poker. We also have in mind more serious "games", such as a contract negotiation between a labor union and a corporation, or an election campaign. Thus, if "game" were not already the established term, "competition" might be more appropriate.

Apart from satisfying our curiosity about clever ways of playing recreational games, game theory has also become the focus of intense interest in studies of business and economics, where the emphasis is on decision-making in a competitive environment. In fact, the language of game theory is becoming more and more a part of main-stream economics. In this book, we will not assume any previous knowledge of economics.

In recent years, game theory has also entered into diverse fields such as biology (evolutionary stability, RNA phage dynamics, viral latency, chromosome segregation subversion in sexual species, *E. coli* mutant proliferation under environmental stress, and population dynamics) and political science (for example, power distribution in legislative bodies). We shall sometimes make reference to those somewhat more speculative applications. In addition, games also are playing a large role in the farthest reaches of axiomatic set theory in studies of large cardinals and projective determinacy, but we shall not venture into that territory.

Now let us become acquainted with a few of the ideas of game theory. Every game has two or more players. The rules of the game specify how the game is to start, namely, which player or players must start the game and what the situation is at the start. That situation is called the *initial* position. Other positions may also occur in the game. At each position, the rules indicate which player (or players) makes a *move* from that position and what are the allowable moves from that position to other positions. At each position p, there must be at least one sequence of moves from the initial position to p. (Otherwise, that position could never enter into the play of the game and would be superfluous.) Some positions are designated as *terminal* positions; no moves are allowed from such a position, so that the game ends when such a position is reached. A play of the game consists of a sequence of moves starting at the initial position and ending at a terminal position. If the rules of the game are such that an infinite sequence of moves is impossible, then the game is said to be *finite*. Chess and checkers, for example, are finite because the rules state that a play of the game must end when the same position has recurred a certain number of times.

Every terminal position determines a *pay-off* to each of the players. In many games, these pay-offs are numbers. Usually these numbers are the amounts of money that the various players win or lose at that terminal position (positive numbers indicating winnings and negative numbers losses). If the sum of the pay-offs at each terminal position is zero, then the game is called a *zero-sum* game. Most recreational and gambling games can be considered zero-sum; whatever some players win has to be balanced by the losses of the other players. On the other hand, many games in economics and politics are not zero-sum; for example, a collective bargaining agreement might result in gains or losses for both sides.

Sometimes, a numerical pay-off may indicate simply whether the player has won or lost. For example, 1 may indicate a win and –1 a loss. In games in which draws (ties) can occur, a draw may be designated by a zero. Alternatively, a win could be shown by W, a loss by L, and a draw by D.

Very simple games can be pictured by a diagram, more precisely by a directed graph. The various positions are represented by nodes (vertices, points) of the graph. Each node, shown as a small open circle, is labeled to indicate which player (or players) must move at that position. A possible move from a given position to another position is represented by an arrow. These arrows can be labeled for easy reference. The node representing the initial position may appear at the top of the diagram or may be specified in some other way. At each terminal node (that is, a node representing a terminal position), the pay-offs will be shown by a suitable n-tuple. For example, if there are three players A, B, and C, then the triple $(2,-1,4)$ at a given terminal node indicates that a play of the game ending at that node yields pay-offs of 2, –1, and 4 to A, B, and C, respectively. (If the game in question ends only in a win (W) or a loss (L) for each player, then the triple (L,L,W) would indicate that A and B have lost and C has won.)

Example 0.1
Consider the following game for two players A and B. Start with a pile of four sticks. At each move, a player can remove either one stick or two sticks. The last player to move wins and the other player loses. Player A has the first move. The directed graph for this game is shown in Figure 0.1. (Each segment is assumed to be an arrow directed downward.) If you are player A, what would you do on your first move?

Until now, we have ignored a large class of important games, those that involve an element of chance. Card games like poker and casino games like roulette or craps always depend on the outcomes of uncertain events, like a spin of a dial or a roulette wheel or the throw of a pair of dice. In order

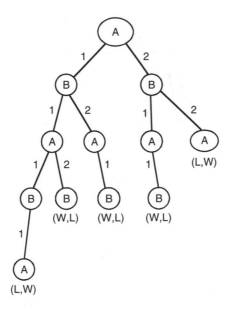

Figure 0.1

to take such games into account, let us allow some positions to be positions at which the choice of one of the possible moves is not made by one of the players but by a device that selects each of the possible moves with a certain probability. For example, the device might toss a coin and then choose one move if the coin shows a Head and another move if the coin shows a Tail. A position that assigns probabilities to the various possible moves from that position is called a *random position* and a move from such a position is called a *random move*. In a diagram of a game, a random position can be labeled with a special symbol, such as "R," and the numerical probabilities of the moves from that position can be attached to the arrows representing those moves.

A game in which there are no random positions is said to be *deterministic*. Thus, the outcome of such a game depends only on the choices made by the players themselves and involves no element of chance. Chess, checkers, and Tic-Tac-Toe are examples of deterministic games.

A very simple example of a two-person nondeterministic game is the following: A fair coin is thrown. Without seeing the result, player B guesses whether the coin shows a Head or a Tail. If B's guess is correct, player A pays B one dollar, whereas, if B's guess is incorrect, B pays A one dollar. The graph of this game is shown in Figure 0.2.

Another characteristic of many games is that the outcome of certain moves may not be known to all of the players. For example, in certain auctions a player's bid is known only to that player. In many card games, the initial random move that consists of a distribution of cards is such that each player only knows the cards that were dealt to that player. Thus, in

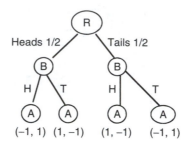

Figure 0.2

such cases, a player whose turn it is to move may not know which of several positions that player may be in.[1] When such situations cannot occur, that is, when the outcome of every possible move is known to all the players, the game is called a *game with perfect information*. Chess, checkers, and Tic-Tac-Toe are examples of games with perfect information.

An example of a game without perfect information is Matching Pennies. In that game, player A puts down a penny, but player B does not see which side is face up. Player B then puts down a penny. If both pennies show Heads or both show Tails, B wins a penny from A. Otherwise, A wins a penny from B. This game is finite, deterministic, and zero-sum.

Exercise 0.1

For each of the following games, determine whether it has the properties of being finite, deterministic, zero-sum, or with perfect information. Describe the initial positions and the terminal positions, if these are not already specified. When feasible, draw a directed graph for the game and indicate the pay-offs at each terminal point.

1. Chess
2. Tic-Tac-Toe
3. The game in Example 0.1 above
4. The first player A names a positive integer n. Then the second player B names a positive integer k. If the sum $n + k$ is even, A wins one dollar from B. If the sum $n + k$ is odd, B wins one dollar from A.
5. Two players A and B alternately name integers from 1 through 9 (repetitions are allowed) until the sum of all the numbers mentioned is greater than or equal to 100. The last player to name a number pays one dollar to the other player.
6. Players A and B, one after the other, show one or two fingers behind their backs, without the other player seeing the opponent's choice. Then A and B guess the number of fingers displayed by their opponent.

---

[1]In the directed graph of the game, those positions at which a player may be located for a particular move are often shown by encircling them by a closed curve.

If both guesses are right or both wrong, the result is a draw. If only one player guesses correctly, then that player wins from the other player, an amount in dollars equal to the sum of the number of fingers displayed by both players.

7. A fair coin is tossed. If a Head turns up, player A pays player B one dollar. If a Tail turns up, B pays A one dollar.

8. (Simple poker) One card is dealt at random to player A from an ordinary deck. Player B does not see what that card is. Player A then either "folds", pays player B one dollar, and the game is over, or A "bets" three dollars. If A bets, then B can either "fold", pay one dollar to A, and the game is over, or B can "call". If B calls, then A wins three dollars from B if A's card is red, or A loses three dollars to B if A's card is black.

9. A fair coin is tossed until a Head turns up. If the number of tosses is even, player A pays player B one dollar. If the number of tosses is odd, B pays A one dollar.

A fundamental notion in game theory is that of *strategy*. The original meaning of the word was "the art of directing the larger military movements and operations of a campaign," but the word has come to be used also in the wider non-military sense of any plan for attaining one's ends. In game theory, we have in mind a special and much more precise meaning. A strategy for a player is a specification that tells the player what that player should do in every situation that might arise during a game. A strategy leaves nothing to the imagination; once a strategy is chosen, the player's moves are completely determined. When all the players choose their strategies, the course of the game and its outcomes are determined; the players could leave the actual performance of the moves to assistants or to machines.

For almost all games, there are so many strategies that the description of all of them is a practical impossibility. In fairly complicated games like checkers or chess, the specification of any useful strategy would be very difficult to formulate and extremely difficult to write down, although something of the sort is done when computers are programmed to play checkers or chess. Good checkers and chess players depend upon their own intuition and experience, not upon an explicit strategy.

Some strategies for a player may be better than others. In a game with numerical pay-offs, a strategy is said to be *non-losing* if the player following that strategy always receives a nonnegative pay-off, no matter what the player's opponents do. Similarly, a strategy is said to be a *winning strategy* if the player following that strategy always receives a positive pay-off, no matter what the player's opponents do. In a game in which Win, Lose, or Draw are the possible pay-offs, a *winning strategy* is one that guarantees a Win, no matter what the opponents do. A *win-or-draw strategy* is simply a non-losing strategy, that is, one that guarantees a Win or a Draw, no matter what the opponents do. In general, it is possible that no player has a winning or a non-losing strategy. (However, in Chapter 1, we shall study a large class of two-person games in which at least one of the players must have a non-losing strategy.)

Example 0.2

Starting with a pile of four sticks, players A and B move alternately, taking away one or two sticks at each move. The last person to move loses. (Notice that, in a certain sense, this is the opposite of the game in Example 0.1, where the last person to move wins.) Player B has the following winning strategy in this game. If A starts by taking one stick, then B should remove two, forcing A to remove the last stick. If A starts by removing two sticks, then B should take away one stick, again forcing A to remove the last stick. In either case, B wins.

Exercise 0.2

Find a winning strategy for one of the players in games (3), (4), and (5) of Exercise 0.1.

Exercise 0.3

In the game of Example 0.1, how many strategies does player A have and how many strategies does B have?

Games in which the pay-offs have numerical values may give a false impression of precision because the real value of the pay-offs may differ significantly from player to player. For example, a pay-off of one hundred dollars to a poor person may be of much greater value to that person than the same pay-off to a millionaire. In order to take this into account, the pay-offs should be adjusted so that they represent their value or utility to the various players. For each player, there would have to be a utility function that transforms numerical pay-offs into numbers that measure the actual value of those pay-offs for that player. This leads to a quagmire of complicated, difficult issues that are not directly connected with game theory itself.[2] For that reason and in order not to make this introductory text needlessly complex right at the beginning, we shall generally sidestep the issue of utility and treat numerical pay-offs in a straightforward way. However, there will be occasions when the issue cannot be ignored.

The book's pedagogical strategy will be to start with relatively simple concepts and theorems. This does not mean that the early going will always be easy. For example, the combinatorial games that form the subject of Chapter 1 are conceptually simple and the basic theorem about them has a simple proof, but the analysis of individual games can involve problems of the highest order of difficulty. In each succeeding chapter, further complexities and more demanding arguments will be encountered, but we shall try to introduce them in small, digestible bites. As much as possible, we shall

---

[2]Helpful discussions may be found in Luce and Raiffa [1957] and in the ground-breaking treatment in von Neumann and Morgenstern [1944]. There is a very brief introduction to an axiomatic treatment of utility theory in Appendix 2.

avoid the use of complicated mathematical concepts and notation, in favor of ordinary language and familiar symbolism.

Chapter 1 deals with certain games, called combinatorial games, that are of general interest. Many of those games are recreational in nature. Readers who are primarily interested in the connections of game theory with business and economics, politics, or biology and other sciences can skim through Chapter 1 and then move on to Chapter 2.

chapter one

Combinatorial games

1.1 Definition of combinatorial game

Combinatorial games constitute a large class of games that includes some familiar games and offers great scope for mathematical ingenuity. At the same time, the relative simplicity of their definition recommends them as our initial entry into the theory of games.

By a *combinatorial game* we shall mean any game with the following characteristics.

1. There are only two players.
2. It is deterministic, that is, there are no random moves.
3. It is finite, that is, every play of the game ends after a finite number of moves.
4. It is a game with perfect information, that is, the result of every move by one player is known to the other player.
5. It is a zero-sum game. (Games that end with a win for one player and a loss for the other player, or a draw for both players, qualify as zero-sum games. This is a consequence of awarding a pay-off of 1 for a win, a pay-off of −1 for a loss, and a pay-off of 0 for a draw.)

Thus, combinatorial games are finite, deterministic, zero-sum, two-person games with perfect information.[1]

A famous example of a combinatorial game is Tic-Tac-Toe (also known as Noughts and Crosses). The game is played on a grid of three rows and three columns. One player uses crosses (X's) and the other player noughts (O's). The players alternately place their own symbols (X or O) in any unoccupied square. The X player moves first. The first player to completely fill a row, column, or either diagonal with that player's symbol wins. Figure 1.1 shows some examples of Tic-Tac-Toe results.

---

[1]The term "combinatorial game" has been used in various ways in game theory. Our definition picks out one way of making it precise.

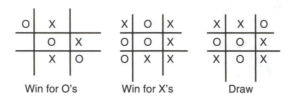

Figure 1.1

Exercise 1.1

Which of the following are combinatorial games?

1. The first player A names a positive integer n. Then the second player
    B names a positive integer k. If the product nk is even, A wins one
    dollar from B; if the product nk is odd, B wins one dollar from A.
2. Removal games.
    a. To start with, there is a pile of 5 sticks. Each of the two players
        alternately removes one or two sticks. The game continues until
        there are no sticks left. The last person to move loses.
    b. The same game as in (a), except that the last person to move wins.
3. The first player A chooses one of the numbers 1 and 2, and then the
    game is over (without player B making any moves at all). A pays B
    one dollar if A has chosen 1, and A pays B two dollars if A has
    chosen 2.
4. Chess.

Recall your childhood experience with Tic-Tac-Toe. You probably will
remember that, after a while, you learned how to make sure that you never
lost, although it is likely that you never explicitly formulated what you had
to do to ensure at least a draw. If your opponent was as experienced as you,
the game would always end in a draw. Hence, the game no longer was of
interest and you stopped playing it. It is an amazing fact that Tic-Tac-Toe is,
in an important way, typical of all combinatorial games. We shall prove later
that, in any combinatorial game, either one player has a winning strategy,
that is, that player can always ensure a win, or both players have non-losing
strategies, that is, they can make sure that they never lose.

Even in as simple a game as Tic-Tac-Toe, the description of a strategy
can be enormously complicated. Therefore, we shall first look at some very
simple games.

Example 1.1

There is a pile of 3 sticks. The players alternately remove 1 or 2 sticks
from the pile. The last person to move loses. Let A be the player who
moves first and let B be the other player. Player A has two strategies (A_1)
and (A_2).

(A$_1$) Remove 1 stick from the pile. (No further specification is required. If player B then removes 1 stick, player A must remove the last stick. If player B removes two sticks, the game is over.)
(A$_2$) Remove two sticks. (Player B must then remove the last stick.)

Player B also has two strategies (B$_1$) and (B$_2$).

(B$_1$) If player A took 1 stick, remove 1 stick. (Player A must then remove the last stick.)
(B$_2$) If player A took 1 stick, remove two sticks. (Then the game is over.)

Notice that player B need not specify what he would do if player A took away 2 sticks. In that case, B would be forced to take away the last stick.

Player A has a winning strategy, (A$_2$). When A removes 2 sticks, B must remove the last stick and B loses. Observe that (A$_1$) is not a winning strategy; if B responds with strategy (B$_1$), that is, if B removes a second stick, then A is left with one stick to remove and loses.

Example 1.2
Consider the same game as in Example 1.1, except that the last person to move *wins* the game. The possible strategies are unchanged. However, player B now has a winning strategy, (B$_2$).

Example 1.3
Consider a game like that of Example 1.2, except that the initial pile consists of 13 sticks. The players move alternately, taking 1 or 2 sticks, and the last person to move wins. (As always, player A moves first.) There are so many strategies for both players that we shall not bother to list them. However, it turns out that player A has the following simple winning strategy. Player A starts off by removing 1 stick, leaving 12. Thereafter, whenever player B removes 1 stick, A follows by removing 2 sticks, and whenever player B removes 2 sticks, A follows by removing 1 stick. Player A must eventually remove the last stick. The reason for this is that, after each of A's moves, the number of sticks left is always a multiple of 3 (that is, 12, 9, 6, 3, and, finally, 0). B will never find a pile consisting of just 1 or 2 sticks and, therefore, B will not be the last person to move.

Example 1.4
Let us describe a non-losing strategy for the first player A, the crosses (X's) player, in the game of Tic-Tac-Toe.

First move: Move into the center.

Case 1. The second player B, the noughts (O's) player, follows by moving into a corner square. By the symmetry of the Tic-Tac-Toe board, we may assume that it is the upper left-hand corner.

Player A should now counter with:

Player B's next move is forced:

O │   │ O
──┼──┼──
  │ X │
──┼──┼──
X │   │

(When we say that a move is *forced*, we mean that any other move would lead to immediate defeat.)

Player A has to reply as follows, to avoid losing:

O │ X │ O
──┼──┼──
  │ X │
──┼──┼──
X │   │

B's next move again is forced:

O │ X │ O
──┼──┼──
  │ X │
──┼──┼──
X │ O │

A continues as follows:

O	X	O
X	X	
X	O	

B is forced to make the following move:

O	X	O
X	X	O
X	O	

A fills the remaining square:

O	X	O
X	X	O
X	O	X

and the result is a draw. Thus, in Case 1, either player A wins (if B fails to make the indicated forced moves) or there is a draw.

Case 2. Player B follows A's first move by moving into a non-corner square. By symmetry, we may assume that B moves into the top row:

	O	
	X	

Player A quickly follows with:

X	O	
	X	

B's next move is forced:

X	O	
	X	
		O

Player A then applies the *coup de grace*:

This yields a double threat along the first column and along one of the diagonals. No matter how B moves, A can win on A's next move. So, in Case 2, player A wins. Thus, in either case, player A will not lose when he plays according to the indicated instructions.

Exercise 1.2
Find winning strategies in the following games.

1. Player A names a positive integer n and player B then names a positive integer k. A wins \$1 from B if the sum $n + k$ is odd, and B wins \$1 from A if the sum $n + k$ is even.
2. The game of Exercise 1.1(1)
3. The same game as Example 1.1 except that the initial pile contains 4 sticks. (Remember that 1 or 2 sticks are removed each time, and the last person to move loses.)
4. The same game as Example 1.1 except that the initial pile contains 5 sticks.
5. Consider the general case of the game in Example 1.1, where the initial pile contains n sticks. Show that the second player B has a winning strategy when n is of the form $3k + 1$ (that is, 1, 4, 7, 10, 13,...), and the first player A has a winning strategy in all other cases.

Exercise 1.3
Consider the general case of the game in Example 1.2, where the initial pile contains n sticks. Recall that the players remove 1 or 2 sticks at a time and the last person to move wins. Describe those values of n for which the second player B has a winning strategy and those values of n for which the first player A has a winning strategy.

Exercise 1.4
Find a non-losing strategy for the second player (the noughts player) in Tic-Tac-Toe. (There are more cases to be handled here than were necessary for the first player's non-losing strategy in Example 1.4.)

Exercise 1.5
Find a winning strategy for the first player in Exercise 0.1(5) of the Introduction.

Exercise 1.6

Find a winning strategy for the first player in Exercise 1.1(3).

1.2 The fundamental theorem for combinatorial games

Fundamental theorem (Zermelo [1912])

In any combinatorial game, at least one of the players has a non-losing strategy. (If draws, that is, zero pay-offs, are impossible, it follows that one of the players has a winning strategy.)

Proof. Let us assume, for the sake of contradiction, that neither the first player A nor the second player B has a non-losing strategy. We shall show that this would imply the possibility of a play of the game with infinitely many moves, contradicting the finiteness condition for combinatorial games.

By a *non-losing position* for player A we mean a situation in the play of the game from which point player A then has a non-losing strategy. Similarly, we define a non-losing position for player B as a situation in the play of the game from which point player B then has a non-losing strategy. By our assumption that neither player has a non-losing strategy for the whole game, it follows that the initial position P_1 of the game is a non-losing position for neither A nor B. In this position P_1, it is A's turn to move.

1. Any move that player A can make from position P_1 cannot lead to a non-losing **position for A**. (If there were such a move M leading to a non-losing position P_2 for A, then the original position P_1 would have been a non-losing position for A. In such a case, A's non-losing strategy starting at P_1 would be to make the move M and then follow the non-losing strategy available for A at the new position P_2.)

2. There is at least one possible move M leading from the original position P_1 to **a new position P_2, which is not a non-losing position for player B**. (First of all, the game cannot end at position P_1. If it did, then, since this is a zero-sum game, at least one of the players would receive a nonnegative pay-off. But then, for that player, position P_1 would automatically be a non-losing position, contradicting our original assumption.[2] Second, if all of the moves open to player A at position P_1 led to non-losing positions for B, then P_1 would be

---

[2]Notice that, in the game in question, we can include among strategies the "null strategy", which simply consists of doing nothing. If a particular game always must end before a player is required to make a move, then that player has a null strategy. If the game never ends with a loss for the player, then the null strategy is a non-losing strategy for that player.

a non-losing position for B, since, no matter how A moved from position P_1, B would be able to use a non-losing strategy for the remainder of the game.)

In accordance with (2), let A make a move M leading from the original position P_1 to a new position P_2 which is not a non-losing position for B. By (1), P_2 is also not a non-losing position for A. Thus, P_2 is a non-losing position for neither A nor B. Now, by repeating the same argument that we have just given, we see that there must be a move from P_2 to a new position P_3 that is again a non-losing position for neither A nor B. In this way, we conclude that there is a possible infinite sequence of moves $P_1 \rightarrow P_2 \rightarrow P_3 \rightarrow ...$, contradicting the finiteness of a combinatorial game.   ∎

Corollary of the fundamental theorem

In a combinatorial game, if one player X has a non-losing strategy, but has no winning strategy, then the other player Y must have a non-losing strategy. Consequently, in a combinatorial game, either one player has a winning strategy or both players have non-losing strategies.

Proof. Let G denote the original combinatorial game. Assume X has a non-losing strategy for G but not a winning strategy. Define a new combinatorial game G* that is the same as G except with respect to certain pay-offs. Whenever the original game G resulted in a draw, let the new game G* assign 1 to player Y and –1 to player X. Thus, G* has no draws. The Fundamental Theorem tells us that X or Y has a non-losing strategy in game G*. Since there are no draws in G*, it follows that X or Y has a winning strategy in G*. Assume that X has a winning strategy in G*. This same strategy would be a winning strategy in the original game G. (For, whenever X receives a positive pay-off in G*, X receives the same pay-off in G.) But this contradicts our hypothesis that X has no winning strategy in G. Therefore, it is impossible for X to have a winning strategy in G*. The only possibility left is that Y has a winning strategy in G*. But such a strategy would be a non-losing strategy for Y in G. For, whenever this strategy does not result in a draw in G, Y receives the same pay-off in G as in G*. But this strategy, being a winning strategy for Y in G*, would yield in such cases a positive pay-off for Y in G*, and hence also in G. Thus, this strategy never results in a loss for Y in the original game G.   ∎

Now consider any combinatorial game. By the Fundamental Theorem, at least one of the players X has a non-losing strategy. If X does not have a winning strategy, then the Corollary tells us that the other player Y also has a non-losing strategy. In that case, if X and Y both use a non-losing strategy, then the game must end in a draw. Thus, there are two possibilities:

1. Exactly one of the players has a winning strategy.
2. Both players have non-losing strategies.

In looser language, we can say that any combinatorial game is either *unfair* (that is, one of the players always will win by correctly playing a winning strategy) or it is *uninteresting* (that is, if both players correctly play their non-losing strategies, a draw always results). *Note that, if draws are impossible in a given combinatorial game, then one of the players must have a winning strategy.*

Our proof of the Fundamental Theorem is only an existence proof. It shows that the non-existence of non-losing strategies for both players is impossible. But it does not tell us how to find such strategies. From a practical point of view, therefore, it is not surprising that many combinatorial games are still unsolved. In the first place, we may not know which of the players has a non-losing strategy. In chess, for example, experience seems to indicate a slight advantage for the first player, White, but it is still conceivable that the second player, Black, might have a non-losing strategy, or even a winning strategy. Computer programs that have been devised to play chess are not based on a winning or a non-losing strategy, but rather use various piece-meal rules culled from the experience of chess experts. Chess is so complex that it may never yield to a definitive analysis.

A second limitation on the application of the Fundamental Theorem is that we may know which player has a non-losing strategy without actually being able to describe such a strategy for practical purposes. Examples of this predicament will turn up when we examine the game of Hex.

Exercise 1.7

Find winning or non-losing strategies in the following games.

1. Starting with a pile of 14 sticks, two players alternately remove 1, 2, or 3 sticks. The last person to move wins. (Hint: Since draws are impossible, one of the players must have a winning strategy. The first player A can ensure that, after each of the second player B's moves, the number of sticks left is not a multiple of 4. Since 0 is a multiple of 4, B cannot remove the last stick. Compare Example 1.3.)
2. Let n be a positive integer. Starting with a pile of n sticks, two players alternately remove 1, 2, or 3 sticks. The last person to move wins. (Hint: The answer depends on n. First consider the case where n is not a multiple of 4.)
3. Let n be a positive integer. Starting with a pile of n sticks, two players alternately remove 1, 2, or 3 sticks. The last person to move loses.
4. Starting with a pile of 14 sticks, two players alternately remove 1, 2, 3, or 4 sticks. The last person to move wins.
5. Let n be a positive integer and let k be an integer such that $1 \le k < n$. Starting with a pile of n sticks, two players alternately remove

any number of sticks between 1 and k. The last person to move wins.

6. The same game as (5) except that the last person to move loses.

Exercise 1.8   *Symmetry*
Find winning or non-losing strategies in the following games.

1. Two players take turns placing pennies on a large table (which is assumed to be circular or rectangular) until no further pennies can be put down. No penny may overlap another penny. The last person to put down a penny wins.
2. An 8×8 checkerboard contains 64 one-inch squares. Two players take turns covering two adjacent squares with 1×2-inch dominoes. The game is played until no further move is possible. The last person to move is the winner.
3. °Same as (1) except that the last person to move loses.[3]
4. °Same as (2) except that the last person to move loses.
5. Consider 25 sticks arranged in a 5×5 square:

$$
\begin{array}{ccccc}
1 & 1 & 1 & 1 & 1 \\
1 & 1 & 1 & 1 & 1 \\
1 & 1 & 1 & 1 & 1 \\
1 & 1 & 1 & 1 & 1 \\
1 & 1 & 1 & 1 & 1 \\
\end{array}
$$

Players alternately take any number of sticks from a single row or column. At least one stick must be taken. There is an additional restriction that a group of sticks cannot be taken if the group contains a gap. (For example, if the second stick of the first row already has been removed, a player cannot, in one move, remove the first and third sticks of the first row.) The last person to move wins.

6. Same as (5) except that there are six rows and six columns.
7. °Same as (5) except that the last person to move loses.
8. Same as (2) on an $n \times n$ checkerboard. (This game is called CRAM. See Cowen-Dickau [1998].)

Exercise 1.9   *Variations and disguised forms of Tic-Tac-Toe*
Find winning or non-losing strategies in the following games.

1. (Martin Gardner) From an ordinary deck of cards, the spades from the ace to the nine (that is, ace, two, three, ..., nine) are placed face up on a table. Two players choose in turn from these cards. A card,

---

[3]The symbol "o" stands for "open" and means that a solution is not known to the author.

once chosen, cannot be chosen again. A winner is the first player to get three cards adding up to fifteen (with the ace counting as one). Hint: Use the following magic square:

2	7	6
9	5	1
4	3	8

Note that, in a magic square, the rows, column, and diagonals all add up to the same number, in this case, fifteen.

2. (Leo Moser) Two players alternately choose a word from the following list of nine words: pony, shin, easy, lust, peck, puma, jury, bred, warn. A player who is the first to have chosen three words having a letter in common wins. The same word cannot be chosen twice.

3. (Opposite Tic-Tac-Toe). This is the same as ordinary Tic-Tac-Toe except that the first person who completes a row, column, or diagonal with that person's marks loses. (Hint: Let the first person move into the center and then copy the opponent's moves symmetrically with respect to the center. The second player's strategy is less obvious.)

4. David and Jill play regular Tic-Tac-Toe. However, in this new game, David wins if the original game is a draw, whereas Jill wins if the original game does not end in a draw. (Hint: Jill has a winning strategy, no matter who goes first.)

5. Regular Tic-Tac-Toe except that neither player can use the center square on the player's first move.

6. Regular Tic-Tac-Toe except that neither player can occupy the center square unless it is a final winning move.

7. Austin's Wild Tic-Tac-Toe (Gardner [1971], Chap. 12) This game is played like ordinary Tic-Tac-Toe except that each player can mark either a cross (X) or a nought (O) in any unoccupied square. The winner is the first player to complete a line (horizontal, vertical, or diagonal) of three marks of the same kind. That player did not have to put in all three marks, only the last one. (Hint: One of the players has a winning strategy.)

8. Ancient Tic-Tac-Toe. In this game, played in ancient China, Greece, and Rome, each player has only three of that player's marks. The players move alternately until all six marks have been placed. Thereafter, the player's alternately move one of their marks to a horizontally or vertically adjacent unoccupied square. The first player to have that player's three marks all on one line (horizontal, vertical, or diagonal) wins. Find a winning strategy for the first player.

9. 4×4 Tic-Tac-Toe.

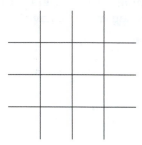

Exercise 1.10
Find winning strategies in the following games.

1. A penny is placed on one of the squares on an edge of a checker-board, and then moved alternately by players A and B. Each move is either forward toward the opposite edge or sideways, one square at a time. It is forbidden to make a move that returns the penny to a position occupied earlier in the game. The winner is the one who pushes the penny into a square on the opposite edge for the first time. Player B places the penny at the start and then player A has the first move.
2. Same as (1) except that the one who pushes the penny into a square on the opposite edge for the first time loses.

Exercise 1.11
Show that, if we extend the notion of a combinatorial game to allow more than two players, then the Fundamental Theorem is no longer true. (Hint: Find a suitable three-person game in which no player has a non-losing strategy.)

We shall examine a few special combinatorial games, the analysis of which is not only far from obvious but also involves the use of various mathematical ideas in quite unexpected ways.

1.3   Nim[4]

At the beginning of a game of Nim, there are one or more piles of objects. Different piles may contain different numbers of objects. The players alternately remove one or more objects from one of the piles (all of the objects in the pile may be taken). The last person to move wins.

It is clear that there are many different forms of Nim, depending upon the number of piles and the number of objects in each pile. Since draws are

[4] The name may be derived from the German word "Nimm" or the archaic English word "Nim", meaning "Take!". The game probably is of Chinese origin. The name Nim was coined by C.L. Bouton, a professor of mathematics at Harvard University, who published in 1901 the first successful analysis of Nim, along the lines which we follow here.

impossible in Nim, one of the players must have a winning strategy. Our analysis will apply to all forms, although which player has a winning strategy will depend upon the particular form of Nim.

Let us look at an especially simple form of Nim:

First pile: 1 1
Second pile: 1 1

Here, the second player B has the following winning strategy.

1. If the first player A takes both objects from a pile, then B removes both objects from the other pile. Then B wins.
2. If A takes one object from a pile, then B takes one object from the other pile.
   We then have:
   First pile: 1
   Second pile: 1
   A must take one object from one of the piles. Then B removes the last object and wins.

Exercise 1.12
Find a winning strategy in the following fairly simple form of Nim:

First pile: 1 1 1
Second pile: 1 1
Third pile: 1 1 1

Our explanation of the winning strategies in Nim will employ a simple arithmetic fact, namely, that every positive integer can be represented in one and only one way as a sum of distinct powers of 2.

Examples

$$1 = 2^0$$
$$2 = 2^1$$
$$3 = 2 + 1 = 2^1 + 2^0$$
$$4 = 2^2$$
$$5 = 4 + 1 = 2^2 + 2^0$$
$$6 = 4 + 2 = 2^2 + 2^1$$
$$7 = 4 + 2 + 1 = 2^2 + 2^1 + 2^0$$

$$\cdots\cdots\cdots$$

$$13 = 8 + 4 + 1 = 2^3 + 2^2 + 2^0$$

$$\cdots\cdots\cdots$$

$$38 = 32 + 4 + 2 = 2^5 + 2^2 + 2^1$$

For each number, the corresponding sum of distinct powers of 2 is called its *binary decomposition*. Thus, the binary decomposition of 22 is $2^4 + 2^2 + 2^1$.

Exercise 1.13
Find the binary decompositions of the numbers from 8 through 12 and the numbers from 14 to 20.

Imagine a finite collection of piles of objects. For each pile, write down the binary decomposition of the number of objects in that pile. The collection of piles is said to be *balanced* if each of the powers of 2 occurs an *even* number of times in the binary decompositions of all the piles; otherwise, the collection is said to be *unbalanced*.

Example 1

First pile:	1 1 1 1	$4 = 2^2$
Second pile:	1 1 1	$3 = 2^1 + 2^0$
Third pile:	1 1 1 1	$4 = 2^2$
Fourth pile:	1 1	$2 = 2^1$

In this case, there are four piles. The binary decompositions have been written so that the same powers of 2 occur in one column. It is clear that this collection is unbalanced, since 2^0 occurs an odd number of times (namely, once).

Example 2

1 1 1	$3 = 2^1 + 2^0$
1 1 1 1 1	$5 = 2^2 + 2^0$
1	$1 = 2^0$
1 1 1	$3 = 2^1 + 2^0$
1 1 1 1	$4 = 2^2$

This collection is balanced. (2^2 occurs twice, 2^1 occurs twice, and 2^0 occurs four times.)

Notice the following facts about balanced and unbalanced collections.

(1) If a player confronts a balanced collection, the new collection **after that player moves must be unbalanced**. This follows from the fact that exactly one pile can be changed in each move. The new binary decomposition for that pile differs from that of the old pile by the presence or absence of at least one power of 2, say 2^j. Since 2^j previously occurred an even number of times in the balanced collection, it must now occur an odd number of times. Hence, the new collection is unbalanced.

(2) If a player confronts an unbalanced collection, that **player can make a move in such a way that the new collection is balanced**. See Examples 3 and 4.

Example 3

Consider the unbalanced collection:

$$1\ 1\ 1 \quad 3 = 2^1 + 2^0$$
$$1\ 1 \qquad 2 = 2^1$$
$$1\ 1\ 1 \quad 3 = 2^1 + 2^0$$

Here, 2^1 occurs an odd number of times. Remove two objects from the first pile, obtaining the balanced collection:

$$1\ 1 = \qquad 2^0$$
$$1\ 1\ 2 = 2^1$$
$$1\ 1\ 1\ 3 = 2^1 + 2^0$$

Example 4

Consider the unbalanced collection:

$$1\ 1\ 1\ 1\ 1 \quad 5 = 2^2 + \qquad 2^0$$
$$1\ 1\ 1 \qquad 3 = \qquad 2^1 + 2^0$$
$$1\ 1\ 1 \qquad 3 = \qquad 2^1 + 2^0$$
$$1\ 1\ 1 \qquad 3 = \qquad 2^1 + 2^0$$

Here, both 2^2 and 2^1 occur an odd number of times. It suffices to take away 2^2 from the first pile and add 2^1 to the first pile. Thus, the new first pile should contain $2^1 + 2^0$ objects. Hence, we should reduce the first pile from 5 objects to 3 objects:

$$1\ 1\ 1 \quad 3 = 2^1 + 2^0$$
$$1\ 1\ 1 \quad 3 = 2^1 + 2^0$$
$$1\ 1\ 1 \quad 3 = 2^1 + 2^0$$
$$1\ 1\ 1 \quad 3 = 2^1 + 2^0$$

The new collection is balanced.

To prove assertion (2) above, consider an unbalanced collection and let 2^k be the highest power of 2 occurring an odd number of times in the binary decompositions of its piles. Find a pile P having 2^k in its binary decomposition. For each power of 2 occurring an odd number of times in the binary decompositions of the piles of the collection, add that power to the binary decomposition of P if it is missing and drop it if it is present. The resulting number is always smaller than the number of objects originally in the pile P, since we dropped 2^k and added at most $2^{k-1} + 2^{k-2} + \cdots + 2 + 1$, which is equal to $2^k - 1$.[5] We can obtain this new number by removing a suitable number of objects from the pile P. The resulting collection is balanced.

[5]Let $S = 2^{k-1} + 2^{k-2} + \cdots + 2 + 1$. Multiplying both sides by 2, we get $2S = 2^k + 2^{k-1} + \cdots + 2^2 + 2$. Now, subtracting the first equation from the second, we obtain $S = 2^k - 1$.

Example 1.5

Consider the unbalanced collection:

1 1 1 1 1 1 1 1	$8 = 2^3$
1 1 1 1 1 1 1 1 1 1	$10 = 2^3 \quad + 2^1$
1 1 1 1 1 1	$6 = \quad 2^2 + 2^1$
1 1 1	$3 = \quad\quad 2^1 + 2^0$

Observe that 2^2 is the highest power of 2 occurring an odd number of times. Choose a pile in which 2^2 occurs. In this case, there is only one such pile, the third pile. Change $2^2 + 2^1$ to 2^0, that is, remove 5 objects from the third pile. (We have dropped 2^2 and 2^1 and we have added 2^0, since these powers of 2 occur an odd number of times in the binary decompositions of the collection.) The new collection is

1 1 1 1 1 1 1 1	$8 = 2^3$
1 1 1 1 1 1 1 1 1 1	$10 = 2^3 + 2^1$
1 1 1 1 1 1	$1 = \quad\quad 2^0$
1 1 1	$3 = \quad 2^1 + 2^0$

This collection is balanced.

By virtue of (1) and (2), it is clear that, if a player X is presented with an unbalanced collection, player X always can ensure that his opponent Y never produces a balanced collection after any of Y's future moves. In fact, X first produces a balanced collection, then Y must turn it into an unbalanced collection, X again can produce a balanced collection, and so on. But this means that X must win. For, after each of Y's moves, the collection is unbalanced and there must be some objects remaining.

Thus, if the original collection is unbalanced, then the first player A has a winning strategy. On the other hand, if the original collection is balanced, then the second player B has a winning strategy. (For, A's first move must produce an unbalanced collection.)

Exercise 1.14

For each of the following forms of Nim, determine which player has a winning strategy. Then practice applying that strategy by playing the game several times.

1. Piles of 3, 5, and 3 objects.
2. Piles of 3, 4, 4, 5, and 6 objects.

Exercise 1.15

You are the first player A in the following game of Nim:

1 1 1 1 1 1 1 1 1
1 1 1 1 1 1 1 1
1 1 1 1 1 1
1 1 1
1 1 1

What should your first move be?

Exercise 1.16

If there are just two piles in Nim, show that the second player has a winning strategy when and only when the two piles contain the same number of objects.

Exercise 1.17

If there are 10 objects distributed among three piles, but you do not know the number of objects in each pile, would you prefer to go first or second in a game of Nim?

Exercise 1.18

Trick based upon binary decompositions.

Peter asks someone to think of any number between 1 and 15, and then to tell him in which of the columns below the number appears.

8	4	2	1
9	5	3	3
10	6	6	5
11	7	7	7
12	12	10	9
13	13	11	11
14	14	14	13
15	15	15	15

Peter adds the numbers at the top of the columns in which the unknown number occurs. The sum is the unknown number. (For example, if the unknown number is 11, Peter will be told that it occurs in the first, third, and fourth columns. Therefore, the sum $8 + 2 + 1$ is the unknown number.)

1. Why does this trick work?
2. How would Peter extend this trick in order to guess any unknown number between 1 and 31?

Exercise 1.19

Prove that every positive integer has one and only one binary decomposition. (Hint: Prove the existence of a binary decomposition by mathematical induction. To prove uniqueness, assume that there are two binary decompositions for the same number and obtain a contradiction by using the formula $2^k + 2^{k-1} + \cdots + 2 + 1 = 2^{k+1} - 1$.)

Reverse Nim

This is the same as Nim except that the last person to move loses. The winning strategy for Nim does not seem to be applicable in Reverse Nim. However, we shall see that it can be modified to yield a winning strategy in Reverse Nim.

Example 1.6

1. Consider the simple case of two piles, each with one object. In Nim, the second player has a winning strategy, whereas the first player has a winning strategy in Reverse Nim. Of course, in both cases, the winning strategy is the only possible strategy.
2. Consider three piles of one object each. In Reverse Nim, the second player has an automatic winning strategy.
3. Consider the game with the following three piles:

    1 1 1
    1
    1

    The first player has a winning strategy in Reverse Nim. Remove two objects from the first pile, reducing the game to that of (2).

Now let us consider an arbitrary game of Reverse Nim. A pile with only one object will be called *unary*. A pile with more than one object will be called *multiple*. We can divide the analysis of Reverse Nim into three cases.

Case 1. All piles are unary. It is obvious that the first player has a winning strategy if and only if there are an even number of piles. The play of the game is automatic.

Case 2. Exactly one pile is multiple. Then the first player has a winning strategy. (Why?)

Case 3. At least two piles are multiple. In this case, we shall show that the player having the winning strategy in ordinary Nim also has a winning strategy in Reverse Nim.

Subcase 3a. Assume that the initial collection of piles is balanced. The second player then has a winning strategy, as in Nim. Player A's first move produces an unbalanced collection. Then B restores a balanced collection, and so on. Eventually, one of the players must leave a collection having only one multiple pile. But a collection with only one multiple pile is necessarily unbalanced. (For only the multiple pile contributes a power of 2 different from 2^0.) Thus, it must be player A who has left a collection with only one multiple pile. It is then B's turn to move and, by Case 2, B has a winning strategy.

Subcase 3b. Assume that the initial collection of piles is unbalanced. Then the first player A has a winning strategy, as in Nim. First, A restores a balanced collection. B then must produce an unbalanced collection, and so on. As in Subcase 3a, the player who first produces a collection with only one multiple pile is the player who always produces unbalanced collections, namely B. Thus, player A eventually confronts a collection of piles with only one multiple pile and, by Case 2, A has a winning strategy.

Exercise 1.20
Determine the winner and the winning strategy in the following examples of Reverse Nim.

1. Piles of 3, 5, and 3 objects.
2. Piles of 3, 4, 2, 5, and 6 objects.
3. Piles of 1, 1, 1, and 2 objects.
4. Piles of 1, 3, 5, and 7 objects. (This form was played in the French movie *Last Year at Marienbad*.)

Exercise 1.21
Analyze the game which is the same as Nim except that the second player B is allowed to dictate the first move of player A.

Exercise 1.22
Analyze the game which is the same as Reverse Nim except that the second player B is allowed to dictate the first move of player A.

1.4 Hex and other games

The combinatorial games discussed so far have either been trivial (Tic-Tac-Toe) or relatively simple to analyze (Nim), or hopelessly complicated (chess or checkers). There is one family of games, however, that combines simplicity and depth in such a way as to make it extremely attractive to mathematicians. This kind of game was invented in 1942 by the Danish engineer, inventor, and poet, Piet Hein. It was marketed in Denmark under the name *Polygon* and was brought out in America under the name *Hex* by Parker Brothers in 1952.

Hex is played on an $n \times n$ board composed of hexagons. For $n = 2, 3, 4, 5$, the boards are shown in Figure 1.2. To describe the boards and moves made on them, we shall number the hexagons on an $n \times n$ board from 1 to n^2, proceeding along each row from left to right, and then down the rows. Once this numbering procedure is adopted, it is not necessary and even

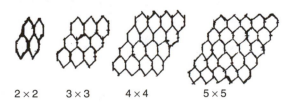

2 × 2   3 × 3   4 × 4   5 × 5

Figure 1.2

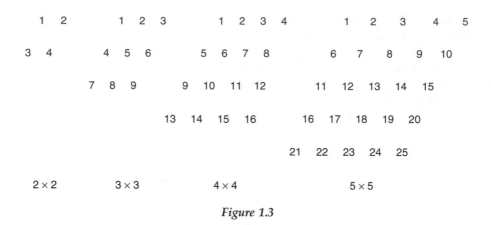

Figure 1.3

easier to omit the hexagons themselves and just write their numbers. Thus, the Hex boards in Figure 1.2 can be drawn as in Figure 1.3.

The opponents are called White and Black, with White moving first. The top and bottom rows are called *White's edges*, and the two sides are called *Black's edges*. (The corners are shared by White and Black.) Thus, White takes the North and South edges, while Black takes the East and West edges. White and Black alternately place their tokens (say, stones or checkers) inside previously unoccupied hexagons. White uses white tokens and Black uses black tokens. White's objective is to construct a continuous chain of white tokens connecting White's edges (that is, the top and bottom of the board), whereas Black tries to construct a continuous chain of black tokens connecting Black's edges (that is, the two sides of the board). Adjacent tokens in a chain must be in hexagons having a common edge. A chain need not be straight. Two examples of winning positions in 5 × 5 Hex are illustrated in Figure 1.4.

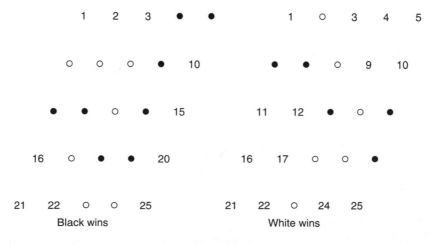

Figure 1.4

In these diagrams, we have omitted the number of a hexagon when it is occupied by a token. The black circles represent Black tokens and the empty circles represent White circles.

Let us look at the simplest forms of Hex.

1. 2 × 2 Hex

    1   2
  3   4

If White moves into hexagon 1, then White will win. For, no matter where Black moves next, White will be able to move into either hexagon 3 or hexagon 4. Thus, White has a winning strategy. (Notice that White also could win by moving into 4 first and then into either 1 or 2, depending on Black's first move. However, if White makes the mistake of moving into 2 or 3 first, then Black can force a win.)

2. 3 × 3 Hex

      1   2   3
    4   5   6
  7   8   9

White again has a winning strategy. First, White moves into 1.

      ○   2   3
    4   5   6
  7   8   9

The rest of the strategy divides into cases, depending upon Black's response.

Case 1.   Assume Black's first move is into any hexagon not in the bottom row. White then moves into either 4 or 5. (At least one of 4 or 5 was still unoccupied.) Say, White moves into 4:

      ○   2   3
    ○   5   6
  7   8   9

(Black's first move is not shown. Black has a token in one of the hexagons 2, 3, 5, 6.) Then, after Black's next move, White will be able to move into either 7 or 8, winning the game.

Case 2.   Assume Black's first move is into 7.

      ○   2   3
    4   5   6
  ●   8   9

Then White moves into 5.

      ○   2   3
    4   ○   6
  ●   8   9

After Black's next move, 8 or 9 is still unoccupied. White moves into 8 or 9, winning the game.

Case 3.   Assume Black's first move is into 9.

```
     o     2     3
   4     5     6
 7     8     •
```

Then White moves into 4.

```
     o     2     3
     o     5     6
 7     8     •
```

After Black's next move, 7 or 8 is still unoccupied. White moves into 7 or 8, winning the game.

Case 4.   Assume Black's first move is into 8.

```
     o     2     3
   4     5     6
 7     •     9
```

White can then force the following sequence of moves, in the sense that, if Black does not make the indicated move, White wins easily in one more move. First, White moves into 9. Black is forced to move into 5. Then White moves into 6. No matter what Black's next move is, White can move into either 2 or 3, winning the game.

Exercise 1.23
1. In 3 × 3 Hex, show that White has winning strategies beginning in any hexagon on the principal diagonal (that is, 1, 5, or 9). We have verified this above for hexagon 1.
2. Show that, in 3 × 3 Hex, White also has winning strategies beginning with hexagons 4 or 6. (The analysis is quite tedious.)
3. Show that, in 3 × 3 Hex, if White's first move is into hexagons 2, 3, 7, or 8, then Black, from that point on, has a winning strategy.

Exercise 1.24
Consider 4 × 4 Hex. (See Figure 1.3.)

1. Show that White has a winning strategy, starting anywhere on the principal diagonal, that is, in any of the hexagons 1, 6, 11, or 16.
2. Show that, if White starts anywhere but on the principal diagonal, Black can ensure a win.

Exercise 1.25
Find a winning strategy for White in 5 × 5 Hex.

To start our general analysis of Hex, we note two facts.

1. It is impossible for both White and Black to win in a game of Hex. This is clear because, if White has a chain connecting the North and South edges, this prevents a chain from being formed between the East and West edges.
2. A draw is impossible in Hex. In other words, after all the hexagons have been filled according to the rules, the only reason for the non-existence of a White North-South chain is the presence of a Black East-West chain. This *seems* intuitively obvious, but we have not proved it. A proof will be outlined later in Exercise 1.29.

The game of Hex was invented independently in 1948 by John Nash, who is responsible for the proof of the following result.

Theorem
White has a winning strategy in Hex.

Proof.   This is an existence proof, that is, we shall show that it is impossible that White does not have a winning strategy, but the proof will provide no way of finding such a strategy. The proof supports our intuitive feeling that White's having the first move certainly couldn't be a disadvantage in a game like Hex, where possession of territory (hexagons) is never a liability.

Assume that White does not have a winning strategy. Since draws are impossible by fact (2) above, the Fundamental Theorem tells us that Black must have a winning strategy. Now, let White proceed as follows. White first makes an arbitrary move, say, in the hexagon 1 in the upper left-hand corner. Thereafter, White pretends that that move was never made. White then follows Black's winning strategy. Since the board apparently does not look the same to White and Black, how does White do this? White looks at the board from the rear (or the other side of the page). So, the board looks to White as it originally appeared to Black, that is, the long diagonal runs from the upper left to the lower right. White then pursues Black's winning strategy (of course, with White and Black reversed). The only thing that can prevent White's winning with this strategy is the possibility that White will be called upon to move into the hexagon 1, into which White made the first move (that White has conveniently forgotten). If there are no other unoccupied hexagons, then, since White is following Black's winning strategy, the move by White into hexagon 1 should win for White. But, since White really has a white token in hexagon 1 from the very beginning, White already has won the game. On the other hand, if there are still unoccupied hexagons different from hexagon 1, White places a white token into one of them, say, hexagon k. White then imagines that White's first move was into k, not into 1. In this way, White is able to continue following this adaptation of Black's winning strategy and eventually wins. So, the assumption that White does not have

a winning strategy implies that White does have a winning strategy. Therefore, White does have a winning strategy.   ∎

Although we now know that White has a winning strategy in $n \times n$ Hex
for any n, we still do not possess such strategies, even for some relatively
small values of n.

Exercise 1.26

By Reverse Hex we mean the same game as Hex except that the first player
to form a continuous chain of his tokens between his edges loses.

   1. Find a winning strategy for White in 2×2 Reverse Hex.
   2. Find a winning strategy for Black in 3×3 Reverse Hex.
   3. Find a winning strategy in $n \times n$ Reverse Hex. Distinguish the cases
       where n is even and n is odd. (The analysis is rather difficult.)

Exercise 1.27

   1. Show that, before the first move in 2×2 Hex, White can place *one* of
       Black's tokens in a suitable position so that White still has a winning
       strategy, but White cannot place *two* black tokens and still have a
       winning strategy.
   2. If we consider the same question as in (1), but this time for 3×3 Hex,
       show that the maximum number of black tokens that White can place
       and still have a winning strategy is *four*.
   3. Solve the analogous problem as in (2) for 4×4 and 5×5 Hex.

Exercise 1.28   Impossibility of a draw in Hex
(Pierce [1961])

Assume that a Hex board has been completely filled in during a game. We
must show that there is either a North-South white chain or an East-West
black chain. Add Western and Eastern black borders and Northern and
Southern white borders, as in Figure 1.5.

Let us construct a path along edges of hexagons in the following manner.
Start off at vertex I in the upper left-hand corner. When we arrive at a vertex,
choose the next edge so that, as we move along it, there is black on the right
and white on the left. (See the two possibilities in Figure 1.5.) Prove that:

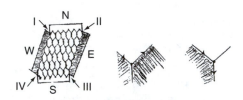

Figure 1.5

1. The path never repeats the same vertex.
2. The path must eventually come to an end.
3. The last vertex in the path does not lie in the interior of the board, that is, it must lie on one of the edges.
4. The last vertex does not lie on the Western edge, except possibly at the lower left vertex IV.
5. The last vertex does not lie on the Southern edge, except possibly at IV.
6. The last vertex does not lie on the Eastern edge, except possibly at the upper right vertex II.
7. The last vertex does not lie on the Northern edge, except possibly at II.
8. If the last vertex is at IV, there is a North-South white chain, and White wins. (Hint: Follow the path from the last time it hits the Northern border until the next time that it hits the Southern border. The White hexagons to the left of the edges of this path form the appropriate chain.)
9. If the last vertex is at II, there is an East-West black chain, and Black wins. (Hint: Follow the path from the last time it hits the Western border until the next time that it hits the Eastern border. The Black hexagons to the right of the edges of this path form the desired chain.)

Exercise 1.29 The Soldiers' game (Schuh [1968],
pp. 239–244; Gardner [1971], Chapter 5) (see Figure 1.6)
The first player A uses three pennies, which start out on positions 1, 2, and 4. Player B uses a nickel, which starts out on position 6. The nickel can move along any line to an adjacent position, but a penny cannot move downward (neither straight down nor downward on a diagonal). The players move alternately. Player A moves only one penny at a time. Player A wins if player B is unable to move. In all other cases, player B wins. This can happen in two ways. Either the nickel can break through past the pennies or a stalemate

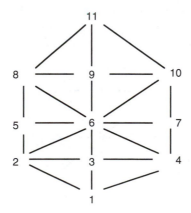

Figure 1.6

can develop. Find a winning strategy for one of the players. (This game was invented during the Franco-Prussian War and was very popular in France.)

Exercise 1.30   The Racks (Silverman [1971])

1. Julia and Josh alternately place the positive integers 1, 2, 3,... into one of the columns I and II.

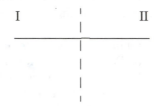

Thus, Julia places 1 in a column, Josh then must place 2 in a column, Julia must then place 3 in a column, and so on. The loser is the first player to put a number w into a column already containing two distinct numbers x and y such that $x + y = w$. In other words, a player must avoid placing a number into a column that already contains two numbers whose sum is the given number. Find a winning strategy for one of the players.

2. The same game as (1) except that there are three columns.

3. Given r columns, what is the largest value of N for which the integers $1, 2,...,N$ can be placed in the columns so that no column contains two numbers together with their sum? (Check that, for $r = 2$, $N = 8$. For $r = 3$, is $N = 23$?)

Exercise 1.31

(Hopscotch) Players White and Black alternately place their tokens on unoccupied positions in Figure 1.7. The object, as in Tic-Tac-Toe, is to place three of your own tokens in a straight line. However, each player is allowed to put down only three tokens. Thereafter, instead of placing more tokens, each player moves one token along a straight line to an adjacent unoccupied position. (Sample game: W into 2, B into 5, W into 6, B into 7, W into 3, B

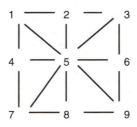

Figure 1.7

into 9, W moves 2 → 1, B moves 5 → 8. B wins.) Show that both players have non-losing strategies.

Exercise 1.32   *Polytechnic*

1. Analyze the following game. Start with a square array of nine dots. Players alternately draw horizontal or vertical lines connecting previously unconnected adjacent dots. When a player draws a line completing a square (or squares), he gets credit for that square (or squares). The player who obtains the largest number of squares wins.
2. Generalize (1) to $n \times n$ arrays of dots.

Exercise 1.33

1. Two players alternately pick integers from 1 to 10 until the sum reaches at least 100. The same number can be used more than once. The first person to reach at least 100 wins. Who has the winning strategy and what is that strategy? Generalize this game to the case where each player picks an integer from 1 to k and the first player to reach n wins.
2. The same game as (1) except that the first person to reach 100 or more loses.

Exercise 1.34

Consider three-dimensional Tic-Tac-Toe. There are 27 boxes and the X's player and the O's player place their markers alternately into unoccupied boxes. The first player to get three of his markers along a straight line (horizontal, vertical, or any of the various "diagonals") wins. Find a winning strategy for the first player.

Exercise 1.35   *Even Wins*

Starting with a collection of 27 objects, the two players alternately remove from 1 to 4 objects at a time until all objects have been removed. Each player then counts the total number of objects that he has removed. The player whose total is even wins. Find a winning strategy for the first player.

Exercise 1.36   *Moore's Nim (Moore [1910])*

1. The game is the same as Nim except that, on any one move, objects can be removed from either or both of any pair of piles.
2. The game is the same as Nim, except that, on any one move, objects can be removed from any group of at most k piles.

Exercise 1.37   *Wythoff's game (Wythoff [1907]; Yaglom and Yaglom [1987], Vol. II, pp. 105–112)*

This is the same game as Nim played with two piles of objects, except that, as an alternative to taking objects from one pile only, a player can take the same number of objects from both piles. The last person to move wins. In the following cases, determine which player has the winning strategy.

1. Piles of 2 objects and 5 objects
2. Piles of 4 objects and 7 objects

1.5   Tree games

A tree game is a game having a diagram of the following kind. At the top of the diagram there is a node for the initial position, from which the first player A makes the first move. The arrows from that position lead to any of a finite number (possibly none) of positions for player B. In turn, the arrows from those positions lead to any of a finite number (possibly none) of positions for player A, and so on. As usual, A and B alternate moves. However, the special condition for a tree game is that no arrow is allowed to lead to any earlier position. Each terminal node at the bottom of the diagram is labeled with A, B, or D, indicating that A wins, B wins, or a draw occurs. An example of a tree game is shown by the diagram in Figure 1.8. Circles indicate a move for A, and squares a move for B. A diagram of a tree game can be thought of as an upside-down picture of a tree; the node at the top is a "root" and the paths leading downward are "branches" of the tree.

Exercise 1.38
In the tree game of Figure 1.8, determine whether one of the players has a winning strategy or whether both players can assure a draw. Hint: Starting at the bottom-most nodes and working upwards, classify each node as either a winning position for A or a winning position for B or a draw position, that is, a non-losing position for both players.

Exercise 1.39
Analyze the tree game of Figure 1.9.

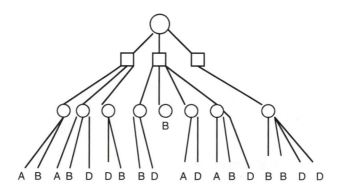

Figure 1.8

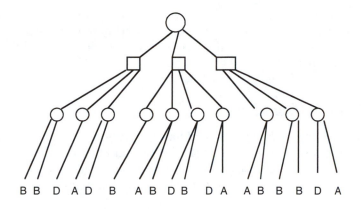

BB   D AD   B   AB DB   DA   AB B   B D A

Figure 1.9

Exercise 1.40
Which player has a winning strategy in the tree game of Figure 1.10?

Exercise 1.41

1. Consider a game in which, starting with five objects, players A and
   B alternately remove one, two, or three objects. The game ends when
   there are no objects left and the winner is the player whose total
   number of objects removed is odd.
   a. Draw the game tree. Use circles for player A and squares for
      player B.
   b. Find out which player has a winning strategy and specify the first
      move of that strategy.
2. Consider a game in which A and B alternately mention 1 or 3 until
   the sum of all the numbers mentioned is at least 6. The last person

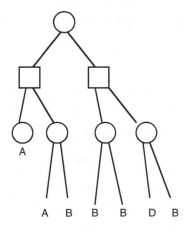

A   B   B   B   D   B

Figure 1.10

to move wins from the other player the final sum of the numbers mentioned. Draw the game tree and find how many strategies each player has. Determine a winning strategy and the amount won by the winner.

1.6   Grundy functions

Sometimes games are so complicated that it is difficult to figure out winning or non-losing strategies. In the case of a special class of games, there is a technique that often enables us to find suitable strategies. The games in question are combinatorial games that have the following two properties:

1.   The last person to move wins.
2.   At each position, there are only a finite number of allowable moves.

Such games are called *Sprague games*.[6] The technique we shall use assigns a nonnegative integer to each position p. That integer will be called the *Grundy number* (or G-number, for short) of p and will be denoted by $g(p)$.[7] A position p will be called *positive* if its G-number $g(p)$ is positive and *zero* if its G-number is 0. To make the definition of the Grundy numbers easier, we let Opt(p) stand for the set of positions that can be reached in one move from p.

The G-numbers are assigned in stages. At the first stage, we assign the number 0 to all positions from which no moves are possible. (Such positions are losing positions.) Thus, if Opt(p) is the empty set $\varnothing$, then $g(p) = 0$. Now assume that we have completed the kth stage of the assignment of G-numbers. If there are positions to which a G-number still has not been assigned, there must be at least one such position p such that all positions in Opt(p) already have G-numbers assigned to them.[8] Then we define the G-number of any such position p to be the least nonnegative integer that is not a G-number of any position in Opt(p). This completes the assignment at the $(k + 1)$th stage.[9]

---

[6]Sometimes these games are referred to as *joli* games.

[7]The technique was discovered independently by the English mathematician Patrick M. Grundy (1917–1959) in Grundy [1939] and the German mathematician Roland P. Sprague (1894–1967) in Sprague [1935, 1936].

[8]If not, an infinite sequence of moves would be possible, contradicting the definition of a combinatorial game. For, from any position p_1 to which no G-number has been assigned, there would be a move to another position p_2 to which no G-number has been assigned. Then, from p_2 there would be a move to a position p_3 to which no G-number has been assigned, and so on. Note that we have used the fact that there is a possible play of the game that reaches position p_1.

[9]Observe that every position has been assigned a G-number. Otherwise, if p_1 were a position to which a G-number was not assigned, it would follow that there is a position p_2 in Opt(p_1) to which a G-number was not assigned. Similarly, there would be a position p_3 in Opt(p_2) to which a G-number was not assigned. Continuing in this way, we could define an infinite sequence of consecutive moves, contradicting the definition of a combinatorial game.

For any finite set F of nonnegative integers, we introduce the notation mex(F) for the least nonnegative integer that does not belong to F. Thus, in the (k + 1)th stage of the definition of the Grundy function G, $g(p)$ = mex $\{g(q): q \in \text{Opt}(p)\}$. The symbolism *mex* comes from the phrase "minimum excluded number."

Example 1.7

Consider a very simple game described by the adjoining graph. The positions are the nodes a, b, c, d, e. An arrow indicates a possible move. The game starts at a. Since Opt(d) = Opt(e) = $\varnothing$, the first stage yields $g(d) = g(e) = 0$. Since Opt(c) = $\{d,e\}$, the second stage yields $g(c)$ = mex $\{0\}$ = 1. Since Opt(b) = $\{c,d\}$, the third stage yields $g(b)$ = mex $\{0,1\}$ = 2. Since Opt(a) = $\{b,c\}$, the fourth stage yields $g(a)$ = mex $\{2,1\}$ = 0. This completes the assignment of G-numbers.

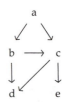

Example 1.8

Consider the game determined by the adjoining graph. Again the game starts at a. Since Opt(f) = $\varnothing$, $g(f)$ = 0. Since Opt(e) = $\{f\}$, $g(e)$ = mex $\{0\}$ = 1. Since Opt(b) = $\{e,f\}$, $g(b)$ = mex $\{0,1\}$ = 2. Since Opt(c) = $\{b,f\}$, $g(c)$ = mex $\{2,0\}$ = 1. Since Opt(d) = $\{c\}$, $g(d)$ = mex $\{1\}$ = 0. Since Opt(a) = $\{b,c,d\}$, $g(a)$ = mex $\{2,1,0\}$ = 3.

The Grundy function furnishes a winning strategy. If the starting position p_1 is positive, the first player A has a winning strategy. For, since $g(p_1)$ = mex $\{g(q): q \in \text{Opt}(p_1)\}$ > 0, there must be some position p_2 in Opt(p_1) for which $g(p_2)$ = 0. Player A should move from p_1 to that zero position p_2. Since $g(p_2)$ = mex $\{g(q): q \in \text{Opt}(p_2)\}$ = 0, either Opt(p_2) = $\varnothing$ and the second player B loses because he cannot make a move, or every move from position p_2 leads to a positive position. In the latter case, player A again can move to a zero position from the positive position that results from B's move. Player A can keep on playing this way and, since the game cannot go on forever and A always has a move available from a positive position to a zero position, B eventually loses. Similarly, if the starting position is a zero position, then B has a winning strategy. For, player A either has no move to make at all or A must move into a positive position and then B has the winning strategy that was just described for A when the starting position was positive.

Let us apply these results to the Sprague games in Examples 1.7 and 1.8 above.

In Example 1.7, the second player B has a winning strategy since the initial position a is a zero position. The first player A can move into b or c. In the former case, B moves into d and wins. In the latter case, B moves into d or e and wins. In Example 1.8, player A has a winning strategy since the initial position a is positive. A should move from a into d, since d is a zero position. B must move from d into c. A then moves from c into f and wins.

Let us look at Wythoff's Game (Exercise 1.37). Recall that we start with two piles of objects. Each move either takes one or more objects from one pile or the move takes the same positive number of objects from both piles. The last player to move wins. Let us analyze the case where we start with at most seven objects in each pile. Consider the 8×8 grid in the following diagram. The entries in the boxes will be explained below.

0	1	2	3	4	5	6	7
1	2	0	4	5	3	7	8
2	0	1	5	3	4	8	6
3	4	5	6	2	0	1	9
4	5	3	2	7	6	9	0
5	3	4	0	6	8	10	1
6	7	8	1	9	10	3	4
7	8	6	9	0	1	4	5

The box in the ith row and jth column (with $0 \le i \le 7$ and $0 \le j \le 7$) represents the position with i objects in the first pile and j objects in the second pile, and the entry in that box is the Grundy number of that position. In fact, an allowable move from a box representing a position in Wythoff's Game goes any number of boxes in the north, west, or northwest directions. (Moving north corresponds to removing objects from the first pile, moving west corresponds to removing objects from the second pile, and moving northwest corresponds to removing the same number of objects from both piles.) The calculation of the Grundy numbers proceeds in the following step-by-step fashion. Since the game ends at $(0,0)$, that position gets Grundy number 0. Since the only move from positions $(1,0)$ or $(0,1)$ is into $(0,0)$, it follows that those positions have Grundy number = mex $\{0\}$ = 1. Since the only boxes to the north, west, or northwest of $(1,1)$ are the three boxes $(1,0)$, $(0,1)$, or $(0,0)$ and they have Grundy numbers 1, 1, and 0, respectively, then $(1,1)$ gets Grundy number = mex $\{0,1\}$ = 2. We can continue in this fashion filling in the Grundy numbers of the rest of the box.

Now we can analyze Wythoff's Game for any initial position in which both piles contain at most seven objects. Consider, for example, the game starting with piles of 6 and 2 objects. Since the entry 8 in the box $(6,2)$ is positive, the first player A has a winning strategy. What should A's first move be? We know that there must be a 0 in a box to the north, west, or northwest of $(6,2)$. In fact, 0 occurs in the box $(1,2)$. Hence, A should remove 5 objects from the first pile.

Exercise 1.42
Find the Grundy numbers of all positions in the game graph below and determine who has a winning strategy and what it is.

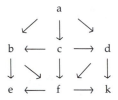

Exercise 1.43
1. In the following initial positions for Wythoff's game, determine which player has a winning strategy and, if that player is A, indicate an appropriate first move for A.
   (a) Piles of 3 and 4 objects
   (b) Piles of 3 and 5 objects
   (c) Piles of 4 and 7 objects
   (d) Piles of 5 and 7 objects
2. Extend the 8 × 8 table above to a suitable 10 × 10 table.

See Vajda ([1992], pp. 51–53), Gardner ([1997], Chapter 8), and Fraenkel [1984] for intriguing facts about Wythoff's Game.

The Grundy function is an example of a function h mapping positions into nonnegative integers such that:

1. $h(p) = 0$ if p is a position for which $\text{Opt}(p) = \varnothing$
2. If $q \in \text{Opt}(p)$, then $h(q) \neq h(p)$
3. If $h(p) > 0$ and $0 \leq k < h(p)$, then there is a position q in $\text{Opt}(p)$ such that $h(q) = k$

Theorem 1.1
Every function h satisfying conditions 1–3 is identical with g.

Proof. Show $h(p) = g(p)$ for all positions p by induction on the stages in the definition of the Grundy function g. By virtue of 1), $g(p) = 0 = h(p)$ for the first stage. Now assume that $h(q) = g(q)$ for all positions up through the nth stage and consider any position p at the $(n + 1)$th stage. Then $h(q) = g(q)$ for all q in $\text{Opt}(p)$. By 2), $h(p) \notin \{h(q): q \in \text{Opt}(p)\}$, and, by 3), all non-negative integers less than $h(p)$ are in $\{h(q): q \in \text{Opt}(p)\}$. Hence, $g(p) = \text{mex } \{g(q): q \in \text{Opt}(p)\} = \text{mex } \{h(q): q \in \text{Opt}(p)\} = h(p)$. This completes the induction. ∎

Now we shall establish a connection between Grundy numbers and the ideas involved in the winning strategy for the game of Nim.

Definition

Given nonnegative integers $n_1, n_2, ..., n_k$, let the *Nim-sum* $n_1 \oplus n_2 \oplus ... \oplus n_k$ be the nonnegative integer N such that a power 2^j appears in the binary decomposition of N if and only if 2^j appears in the binary decomposition of an odd number of the integers $n_1, n_2, ..., n_k$.

Example 1.9
1. $5 \oplus 13$
   $5 = 1 + 4$, $13 = 1 + 4 + 8$. So, $5 \oplus 13 = 8$.
2. $n \oplus n = 0$ for all n
3. $n \oplus 0 = n$ for all n
4. $12 \oplus 7 \oplus 4$
   $12 = 4 + 8$, $7 = 1 + 2 + 4$, $4 = 4$. So, $12 \oplus 7 \oplus 4 = 1 + 2 + 4 + 8 = 15$.

To simplify the computation of a Nim-sum, one can eliminate pairs of powers of 2 that occur in the binary decompositions of the summands. Note that the Nim-sum of the numbers of the piles of a balanced position in Nim is 0, and the Nim-sum of the numbers of the piles of an unbalanced position in Nim is positive.

Corollary 1.2

The Grundy number of a position in Nim is the Nim-sum of the cardinal numbers of the piles in the position.

Proof. It is easy to show that the Nim-sum $h(p)$ of a position p in Nim satisfies conditions 1–3 above. Hence, by Theorem 1.1, $g(p) = h(p)$.   ∎

Example 1.10

Consider the position in Nim with piles of 3, 5, 7, and 9 objects. Since $3 = 1 + 2$, $5 = 1 + 4$, $7 = 1 + 2 + 4$, and $9 = 1 + 8$, the Grundy number of this position is the Nim-sum $3 \oplus 5 \oplus 7 \oplus 9 = 8$.

Observe that Corollary 1.2 yields our earlier method of determining winning positions in Nim. In fact, a balanced position has Grundy number 0 and, therefore, is a losing position, whereas an unbalanced position has positive Grundy number and, therefore, is a winning position.

1.7   Bogus Nim-sums

We shall see that Grundy functions and Nim-sums can be combined in novel ways to find successful analyses of various Sprague games.

Example 1.11   Tape Dominos (Dawson's Kayles)

Consider a tape divided into cells of equal size. Let a move consist of placing a domino on two adjacent cells. No domino is allowed to overlap any previously placed domino. For example, after two moves a 9-cell tape might look like this:

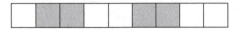

If the original number of cells is even, there is a simple winning strategy for the first player: Place the first domino on the middle two cells and then play symmetrically with respect to the center of the tape. However, when the original number of cells is odd, there does not seem to be a simple winning strategy. A move in this game, consisting of placing a domino on the tape, can be thought of as dividing the game into a *disjoint sum* $G_1 + G_2$ of two games G_1 and G_2. G_1 is the game of tape dominos played on the tape to the left of the domino that was just placed on the tape, and G_2 is the game of tape dominos played on the tape to the right of that domino. Any further move must be a move in either G_1 or G_2.

In general, given games $G_1, ..., G_k$, we define their *disjoint sum* $G_1 \oplus ... \oplus G_k$ to be the simultaneous playing of the component games in the following way: each move consists of a choice of one of the games G_j and a move from the present position in G_j. Hence, the disjoint sum is finished when and only when each of the component games is finished. If p is a position in the disjoint sum, then each component G_j is in a position p_j. That position p_j has Grundy number $g_j(p_j)$ in G_j. Then the Grundy number of p in the disjoint sum is the Nim-sum $g_1(p_1) \oplus ... \oplus g_k(p_k)$. (This follows from Theorem 1.1, since it is not hard to see that the function $h(p) = g_1(p_1) \oplus ... \oplus g_k(p_k)$ satisfies conditions 1–3 of the theorem.) As a consequence, we can pretend that each of the components G_j is a Nim pile containing $g_j(p_j)$ objects. It follows that the winning strategy for Nim can be applied to these "bogus" Nim piles.

Let us see how this Bogus Nim tactic applies to Tape Dominos. Since we eventually have to know the Grundy numbers for a completely blank tape, let us compute the Grundy numbers b_n for a blank tape with n cells. Clearly, $b_0 = b_1 = 0$ (since no move is possible from a tape with no cells or a tape with one cell) and then $b_2 = 1$ and $b_3 = 1$. Further simple computation yields $b_4 = 2$, $b_5 = 0$, $b_6 = 3$, $b_7 = 1$, $b_8 = 1$, $b_9 = 0$, and $b_{10} = 3$. The calculation of additional values requires just some tedious labor.

Example 1.12

Let us find winning moves (if any) for the following Tape Dominos positions.

1.

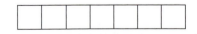

Since $b_7 = 1 > 0$, we know that there is a winning move. In fact, we can obtain

with Grundy number $b_5 = 0$. There are other winning moves, for example, to

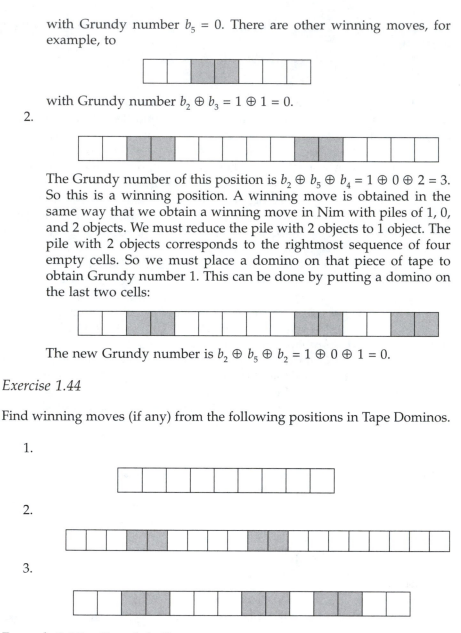

with Grundy number $b_2 \oplus b_3 = 1 \oplus 1 = 0$.

2.

The Grundy number of this position is $b_2 \oplus b_5 \oplus b_4 = 1 \oplus 0 \oplus 2 = 3$. So this is a winning position. A winning move is obtained in the same way that we obtain a winning move in Nim with piles of 1, 0, and 2 objects. We must reduce the pile with 2 objects to 1 object. The pile with 2 objects corresponds to the rightmost sequence of four empty cells. So we must place a domino on that piece of tape to obtain Grundy number 1. This can be done by putting a domino on the last two cells:

The new Grundy number is $b_2 \oplus b_5 \oplus b_2 = 1 \oplus 0 \oplus 1 = 0$.

Exercise 1.44

Find winning moves (if any) from the following positions in Tape Dominos.

1.

2.

3.

Example 1.13   Grundy's Game

Start with piles of objects. Each move must divide a pile into two nonempty piles of *unequal* size.

Since piles of 1 or 2 objects can't be divided, a collection of piles consisting of 1 or 2 objects is a final losing position with Grundy number 0. In order to apply the Bogus Nim technique, we must have available the Grundy number c_n for a single pile of n objects. Clearly, $c_0 = c_1 = c_2 = 0$. Since the only move from a pile of three objects leads to a collection consisting of a one-object pile and a two-object pile, with Grundy number 0, $c_3 = 1$. Further calculation yields $c_4 = 0$, $c_5 = 2$, $c_6 = 1$, $c_7 = 0$, $c_8 = 2$, $c_9 = 1$, and $c_{10} = 0$. Let us analyze the following Grundy games.

1. Piles of 2, 4, and 6 objects. This will have Grundy number $c_2 \oplus c_4 \oplus c_6 = 0 \oplus 0 \oplus 1 = 1$. Since this is positive, the first player has a winning strategy and the first move of that strategy must reduce the third pile to a collection with Grundy number 0. This can be done by dividing it into piles with 2 and 4 objects, with Grundy number $c_2 \oplus c_4 = 0 \oplus 0 = 0$.
2. Piles of 7, 5, and 9 objects. This has Grundy number $c_7 \oplus c_5 \oplus c_9 = 0 \oplus 2 \oplus 1 = 3$. We must reduce this to a Nim-sum 0 by changing the 2 to a 1. So, the second pile with 5 objects must be divided into two piles with Nim-sum 1. Since $c_2 = 0$ and $c_3 = 1$, the first player must divide the second pile into piles of 2 and 3 objects.

Exercise 1.45
Find winning first moves (if any) for the following Grundy games.

1. Piles of 4, 6, and 8 objects
2. Piles of 4, 5, and 8 objects
3. Piles of 3, 5, 5, and 8 objects

Exercise 1.46
Calculate the numbers c_n for $n = 11, 12, \ldots, 20$. Then find a winning first move for the Grundy game with piles of 13, 14, and 18 objects

Exercise 1.47
Consider the game that is just like Grundy's game except that the two piles into which a pile is divided need not be of unequal size.

1. Find a formula for the Grundy number of a single pile of n objects.
2. Show that a position is a winning position if and only if it contains an odd number of even-numbered piles.

Exercise 1.48
Consider a modified Nim in which one can take at most two objects from a pile. Find a formula for the Grundy number of a single pile of n objects.

Nimble

The game of Nimble is played on a tape divided into cells, numbered in increasing order from left to right. Chips occupy some of the cells, with possibly more than one chip in each cell. Each move consists of moving one chip to a cell further to the left, or even off the tape. The chip can jump over other chips or move into the same cell as other chips. The last player to move wins. Nimble is a disguised form of Nim. Consider each chip a Nim pile with the number of objects in the pile equal to the chip's position on the tape.

Exercise 1.49
Find a winning first move from the position on the tape above.

Northcott's Game

This game is played on a checkerboard. There are two chips in each row (but not on the same square), one chip (white) for player A and one chip (black) for player B. At each move, the player moves that player's chip in one of the rows, without jumping over or on top of the other chip in the row. The last player to move wins. This is another disguised form of Nim. The number of squares between opposing chips in a row is a Nim pile. The analogy with Nim is not complete. A player can increase the gap between chips. But then the opponent can restore the original gap and push the other player further toward the edge. Note also that this is, from a strict point of view, not a combinatorial game, since infinite games are possible. Nevertheless, the standard Nim technique works.

Exercise 1.50
Find a winning first move from the following position in Northcott's game.

Enimga (sic)

We start with black chips in cells on a tape, with at most one chip in each cell. A move consists of moving a chip any number of cells to the left without jumping or moving on top of other chips. The last player to move wins.

This game is more difficult to analyze, but it yields to a more subtle Nim-like strategy. List the gaps between chips, starting from the rightmost chip, and including the gap between the leftmost chip and the left border. Count a zero gap between adjacent chips. In the example above, we obtain 3 0 0 0 1 1. Take alternate gaps as Nim piles. In the example, we get 3 0 1. This is unbalanced. Change the 3-pile to a 1-pile by moving the rightmost chip two cells to the left. Ultimately the alternate gaps will all be 0. So the chips will occur in adjacent pairs (and possibly also a chip in the leftmost cell). Then play a chasing strategy; your opponent must move a left chip of one of the adjacent pairs and you then close the remaining gap. Eventually your opponent will be unable to move.

Exercise 1.51
Find a winning first move from the following Enimga position.

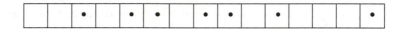

Lasker's Nim

This is the same as Nim except that, in addition to moves in which one pile is reduced, a move can divide a pile into two nonempty piles. The last player to move wins.

Calculation yields the following Grundy numbers for single piles with n objects.

n	$g(n)$	n	$g(n)$	n	$g(n)$
0	0	5	5	9	9
1	1	6	6	10	10
2	2	7	8	11	12
3	4	8	7	12	11
4	3				

One can prove by induction that, for any $n > 0$, $g(4n - 3) = 4n - 3$, $g(4n - 2) = 4n - 2$, $g(4n - 1) = 4n$, $g(4n) = 4n - 1$

Example 1.14
Consider piles of 3, 5, 7, and 10 in Lasker's Nim. The corresponding Grundy numbers are 4, 5, 8, 10, with Nim-sum $4 \oplus 5 \oplus 8 \oplus 10 = 3$. So we must change the Grundy number 10 to 9. This can be done by removing one object from the 10-pile.

Exercise 1.52
Find an initial winning move for Lasker's Nim with piles of 7, 9, and 11 objects.

Subtraction games $S(s_1, s_2, \ldots, s_k)$

Start with a pile of objects. Each move removes a certain number of objects from the pile, where the number comes from a given collection of numbers $s_1, s_2, \ldots, s_k$. The last person able to move wins. The game is denoted $S(s_1, s_2, \ldots, s_k)$.

Example 1.15

1. $S(1,2)$. The players remove one or two objects at each turn. The Grundy function has period 0, 1, 2, that is, the values $g(n)$ of the Grundy function start out with 0, 1, 2, and then keep repeating. As in Exercise 1.2(5), the first player A has a winning strategy when and only when the initial number n of objects is not divisible by 3.

2. $S(1,2, \ldots,k)$. This is a generalization of example (1). The Grundy function has period 0, 1,...,k. Therefore, player A has the winning strategy when and only when the initial number n of objects is not divisible by $k + 1$.

Exercise 1.53

1. In the game $S(s_1, s_2, \ldots, s_k)$, show that $g(n) \le k$ and, therefore, the values $g(n)$ will eventually be periodic.
    (Hint: $g(n) = \text{mex } \{g(n - s_1), \ldots, g(n - s_k)\}$.)
2. Show that the game $S(k)$ has the period 0, ...,0, 1, ...,1, consisting of k 0's followed by k 1's.
3. Find the periods for $S(1,2k + 1)$ and $S(1,2k)$.
4. Find the period for $S(k,k + 1)$.
5. Show that the sequence of Grundy function values for $S(us_1, \ldots, us_k)$ is that of $S(s_1, s_2, \ldots, s_k)$ with each of the entries of the latter repeated u times. Look for example at $S(1,2)$ and $S(3,6)$. Write the period for $S(2,6)$, using the period for $S(1,3)$.
6. Find the periods for $S(2,3,5)$ and $S(1,3,4)$.
7. If b and c are relatively prime, that is, they have no common factor bigger than 1, what can be said about the period of $S(b, c)$?

Exercise 1.54   Kayles (Dudeney [1907])

Begin with a tape divided into n equal squares. At each move, the player can remove one square or two adjacent squares ("adjacent" on the original tape). The last person to move wins.

1. Show that there is a simple winning "symmetry" strategy for one of the players. (Hint: The strategy is different for even and odd n.)
2. Investigate the Grundy function $g(n)$ for Kayles. (Note: It turns out to be periodic for $k \ge 71$ with a period of length 12 consisting of 412814721827, but there are earlier 'partial' periodicities. See Berlekamp et al. [1992], Vol. 1, p. 91.)

Exercise 1.55 Treblecross

On a tape divided into n equal squares, with $n \geq 3$, at each move the player places a black chip in an unoccupied square. The first player to complete a sequence of three consecutive chips wins. Investigate the winning strategies. (See Berlekamp et al. [1992], Vol. 1, pp. 93–94.)

Exercise 1.56 Odd Wins

Given a pile of n objects, with n odd. Each move takes away at least one and up to k objects. The winner is the player whose final total of objects removed by that player is odd. (a) When $k = n - 2$, find a winning strategy for the second player B. (b) No general analysis is known. Find more general solutions than (a).

Exercise 1.57 Odd Nim

This is the same as Nim except that an odd number of objects has to be removed from a pile. (a) Show that the Grundy function g satisfies the conditions that $g(p) = 1$ when p is a position consisting of one odd pile and $g(p) = 0$ when p is a position consisting of one even pile. (b) Show that $g(p) = 1$ when p is a position consisting of an odd number of odd piles and, otherwise, $g(p) = 0$. Hence, the first player has the winning strategy when and only when the initial position contains an odd number of odd piles.

There are a host of other interesting and challenging combinatorial games, many of them still unsolved. Examples may be found in the following references: Averbach and Chein [1980]; Beasley [1989]; Berlekamp et al. [1992]; Conway [1976]; Cornelius and Parr [1991]; Dudeney [1958]; Silverman [1971]; and Vajda [1992]. The books Berlekamp et al. [1992] and Conway [1976] present a more systematic way of classifying and studying combinatorial games.

Here are a few more combinatorial games.

Exercise 1.58 Welter

A tape is divided into n squares, numbered from 1 to n. Some Black chips are placed on some of the squares, no two on the same square. A move consists of moving a chip to a lower-numbered unoccupied square. The last player to move wins.

1. Find a winning strategy for one of the players in the following Welter game.

2. Find a winning strategy for one of the players in the following Welter game.

3. Find a general approach to Welter games (Conway [1976], Chapter 13).

Exercise 1.59

On a 3×3 checkerboard, at each move the player places one cross in one or more unoccupied squares, as long as they are all in the same row or column. The last person to move wins (at which time all squares are filled). Show that the second player B has a winning strategy (Averbach and Chein [1980], pp. 238–245).

Exercise 1.60   Aliquot (Silverman [1971])

Start with an even positive integer. At each move, the player must subtract from the current integer k some positive divisor of k different from k. The last person to move wins. Find a general winning strategy for one of the players.

Exercise 1.61   Subtract-a-square (Silverman [1971])

Start with a positive integer n. The players remove a positive perfect square (1, 4, 9, …) at each move. The last player to move wins.

1. Develop the Grundy function $g(n)$ and find its period.
2. If $n = 44$, find a winning first move.

Exercise 1.62   Gnim (Gale's Nim) (Gale [1974])

Consider mn objects laid out in a rectangular array of m rows and n columns. Let (i,j) stand for the object in the ith row and jth column. At each move, the player chooses some (i,j) not already removed and then removes all (r,s) with $r \geq i$ and $s \geq j$ (that is, the "northeastern" corner with (i,j) at its lower left end). The last person to move loses.

1. Find a winning strategy in $2 \times n$ Gnim (or, symmetrically in $m \times 2$ Gnim).
2. Find a winning strategy in $n \times n$ Gnim.
3. Show that the first player has a winning strategy in $m \times n$ Gnim. (Note: The proof may not show how to find the winning strategy.)

Exercise 1.63   Dudeney's 37 game (Cornelius and Parr [1991], p. 16)

This game is played on the following tape:

1	2	3	4	5

On the first move, the first player A places a chip on one of the cells. At each move thereafter, the player whose turn it is moves the chip to a different cell and adds the number in that cell to the sum of the numbers in all the previously chosen cells. The winner is the first player to make a move yielding a sum of 37 or to force the opposing player to make a move yielding a sum greater than 37. Find a winning strategy.

Exercise 1.64   The Hypergame Paradox (Zwicker [1987])
Let Hypergame be the following game. Player A's first move is to choose a specific combinatorial game G. Then the players go on to play the game G, with player B making the first move of G.

1. Verify that Hypergame is a combinatorial game.
2. Show that an infinite sequence of moves is possible in Hypergame and that, as a consequence, Hypergame is not a combinatorial game.
3. Try to resolve the contradiction between (1) and (2).

Exercise 1.65
The game is played on the following two tapes.

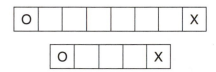

The O-player must move at each of his turns one of the two O's any number of cells to the left or right on the same tape. Likewise, the X-player must move at each of his turns one of the two X's any number of cells to the left or right on the same tape. Two symbols cannot occupy the same cell and no symbol is allowed to jump over another symbol. The O-player is given the first move. The last player to move wins. (Thus, the object of the game is to trap your opponent's symbols.) Find a winning strategy for one of the players.

Exercise 1.66 (Silverman [1971])
On a Tic-Tac-Toe board, each of two players in turn must cross off a positive number of squares in any one row or column. (The same square cannot be crossed off more than once.) The last person to move wins. Find a winning strategy for one of the players.

Exercise 1.67   SIM
1. On a diagram consisting of the six vertices of a hexagon, each of two players in turn must draw a straight line connecting two vertices that have not already been connected. (There are fifteen such lines.) The first player uses a red pencil and the second player a blue pencil. A line cannot be drawn if it completes a triangle all of whose sides have

the same color. The last person to move wins. Show that the game cannot end in a draw and find a winning strategy for one of the players. (This game was invented by Gustavus J. Simmons.)

2. Study the game of Reverse SIM, in which the last person to move loses.

3. What happens in the game of Part (a) when both players use the same color?

Exercise 1.68   Regulus (Silverman [1971])

1. On a diagram consisting of the n vertices of a regular n-gon, each of two players in turn must draw a straight line connecting two vertices. No vertex and no edge may be used more than once, and no two line segments are allowed to intersect. The last player to move wins. Find a winning strategy for one of the players. (You may want to distinguish the cases where n is even and where n is odd.)

2. Study the game of Reverse Regulus, in which the last person to move loses.

chapter two

Two-person zero-sum games

2.1 Games in normal form

In this chapter, we shall confine our attention to zero-sum games with two players. In such games, the competitive aspect is extreme, since whatever is won by one player is lost by the other. We shall impose the important restriction that each player has only a finite number of strategies.[1] Random moves are permitted, so that a game might not be deterministic. Moreover, there is no requirement of perfect information, that is, a player does not necessarily know what moves the opponent has made. In fact, the details of the play of the game ordinarily will not enter into the analysis of the game. The only information required will be a specification of the strategies that are available to the players and the pay-offs that correspond to each choice of strategies.

Assume that player A has strategies $A_1, ..., A_m$ and that player B has strategies $B_1, ..., B_n$. Then, for every pair of strategies (A_i, B_j), there will be a pay-off $P(A_i, B_j)$ for player A. Since the game is zero-sum, the corresponding pay-off for player B must be $-P(A_i, B_j)$. Thus, the game may be described by the $m \times n$ matrix of real numbers in which the entry in the ith row and jth column is $P(A_i, B_j)$. This matrix is called the *pay-off matrix for A*. When a game is specified in this way, the game is said to be in *normal form*.[2]

In a game that contains random moves, a pay-off $P(A_i, B_j)$ actually will be the *expected pay-off*, in the sense of probability theory. A review of elementary probability theory may be found in Appendix 1. According to the frequency interpretation of probability, the expected pay-off $P(A_i, B_j)$ will be the

---

[1] For the sake of mathematical simplicity, we shall ignore the very significant areas of game theory in which a player may have denumerably many or a continuum of strategies. Therefore, every move in a game must have only a finite number of outcomes. For more general treatments, see Burger [1963] or Dresher [1981]. In the denumerable case, infinite series would arise, and, in the continuum case, definite integrals.

[2] In contrast, if a detailed description is given of the possible moves that can occur in a game, then the game is said to be given in *extensive form*. Many examples of such games were given in Chapter 1.

long-run average[3] of the pay-offs when the game is repeatedly played with player A using strategy A_i and player B using strategy B_j.

Example 2.1

A frequently cited example is that of the Battle of the Bismarck Sea. In 1943, during the fighting in and around New Guinea in World War II, a Japanese supply convoy was preparing to sail from Rabaul to Lae. The convoy could use either a northern route with poor visibility or a southern route with clear weather. The American air force knew that the convoy was about to sail. The air force could begin its reconnaissance either in the north or the south. The number of days in which the air force could bomb the convoy depended upon whether the Americans guessed correctly, and was given by the following table:

	North$_J$	South$_J$
North$_A$	2	2
South$_A$	1	3

The rows give the two possible American strategies, and the columns represent the two possible Japanese strategies. The numerical entries give the number of days of bombing. For example, the entry "1" in the South$_A$ row and North$_J$ column indicates that, if the Americans began to look for the convoy in the south and the convoy took the northern route, then there would be only one day of bombing. The estimates of the American and Japanese commanders actually agreed on these figures. Thus, the confrontation between the Americans and Japanese could be thought of as a two-person zero-sum game in normal form with the 2×2 matrix:

$$\begin{pmatrix} 2 & 2 \\ 1 & 3 \end{pmatrix}$$

Example 2.2

A pair of fair dice are thrown and each of players A and B can bet either $12 or $24 on the outcome. Player A wins if the sum of the dice is more than 6; otherwise, B wins. The winner receives from the loser the sum of the two bets. This defines a two-person zero- sum (non-deterministic) game, given in the following normal form. Note that each player has two strategies, Bet $12 or Bet $24. Remember that A's strategies are listed in the rows and B's strategies in the columns.[4]

	B Bets $12	B Bets $24
A Bets $12	4	6
A Bets $24	6	8

[3]That is, the limit of the average of the pay-offs $P(A_i, B_j)$ as the number of games played approaches infinity.
[4]Books on game theory sometimes refer to player A as R (for Rows) and to player B as C (for Columns), since the rows of the game matrix correspond to the strategies for A and the columns of the matrix correspond to the strategies for B.

The matrix of this game is

$$\begin{pmatrix} 4 & 6 \\ 6 & 8 \end{pmatrix}$$

The entries are the expected winnings for player A. For example, assume that both players bet $12. Since a throw of the pair of fair dice can yield 36 equally probable outcomes and 21 of them have a sum greater than 6 (just count the possibilities!), the probability that A wins is 21/36, that is, 7/12. So, either A can win $24 (the sum of the two bets) with probability 7/12, or A can lose $24 with probability 5/12. Hence, A's expected winnings[5] are 24(7/12) + (−24)(5/12) = 4. Next, assume that A bets $12 and B bets $24. Then either A can win $36 (the sum of the bets) with probability 7/12 or A loses $36 with probability 5/12. So, A's expected winnings would be 36(7/12) + (−36)(5/12) = 6. This is the entry in the first row, second column. The other two entries can be calculated in a similar way. Note that the game is biased in favor of A.

Games are usually described in ordinary English. The construction of the corresponding normal form matrix is often straightforward, but may sometimes be rather difficult.

Exercise 2.1
Find the normal form matrix for each of the following games. Remember that the entries are the pay-offs for player A, the row player.

1. Each of players A and B chooses one of the numbers 1 and 2, without either player knowing in advance what the other player will choose. If the sum of the chosen numbers is even, player A receives that sum in dollars from B. If the sum is odd, B receives from A as many dollars as the number B selected.
2. Each of players A and B independently selects Head or Tail, and then a fair coin is tossed. If both players choose the side of the coin that appeared or both choose the side that did not appear, they each receive zero dollars. Otherwise, the player whose choice was correct, receives one dollar from the other player.
3. Player A chooses one of the numbers 1, 2, 3 and player B chooses either 1 or 2. If the sum of the chosen numbers is prime, A receives that sum in dollars from B. If the sum is not prime, B receives that sum from A.

---

[5]In general, if the possible winnings of player A are $x_1, \ldots x_n$ and the probability that A wins x_i is p_i, then the expected winnings of A are $p_1 x_1 + \cdots + p_n x_n$. See Appendix 1.

2.2   *Saddle points and equilibrium pairs*

Now let us see whether examination of the matrix of a game in normal form can produce any worthwhile advice to the players as to how they should play. In fact, there is at least one case in which such advice is available. By a *saddle point* of a matrix we mean an entry that is the minimum in its row and the maximum in its column.[6] For example, in the matrix of the game in Example 2.1, the entry "2" in the first row and first column is a saddle point, and, in the matrix of the game in Example 2.2, the entry "6" in the second row and first column is a saddle point. In a matrix of a game, a pair (A_i, B_j) of strategies for which the corresponding pay-off is a saddle point is called an *equilibrium pair* of the game. Thus, in the game of Example 2.1, the pair (North$_A$, North$_J$) is an equilibrium pair and, in the game of Example 2.2, the pair (A Bets \$24, B Bets \$12) is an equilibrium pair.

An equilibrium pair (A_i, B_j) is stable in the sense that deviation by one of the players is inadvisable. If a player X plays his strategy in that equilibrium pair, then, if the other player Y does not play his strategy in that equilibrium pair, then that other player Y runs the risk of getting a poorer pay-off than he would get by playing his strategy in the equilibrium pair. For example, if (A_i, B_j) is an equilibrium pair and if player B decides to play a strategy B_k different from B_j, then, since $P(A_i, B_j)$ is a saddle point, $P(A_i, B_j)$ is the minimum pay-off in its row; hence, $P(A_i, B_j) \leq P(A_i, B_k)$, so that the new pay-off for player A will be at least as great and possibly greater than the pay-off for A at the equilibrium pair. Similarly, if player A deviated from his equilibrium strategy, his pay-off may decrease because $P(A_i, B_j)$ is the maximum pay-off in its column.

To illustrate this stability of equilibrium pairs, consider first the game in Example 2.1. (North$_A$, North$_J$) is an equilibrium pair. If the Americans changed their strategy to South$_A$, then they would be worse off since the number of bombing days would decrease from 2 to 1. If the Japanese changed their strategy from North$_J$ to South$_J$, then they would gain no advantage since the number of bombing days would remain at 2. In actuality, both sides did choose their equilibrium strategies.

In the game in Example 2.2, the equilibrium pair is (A Bets \$24, B Bets \$12), the second row and first column, with pay-off to A of \$6. If player B plays his equilibrium strategy but player A changes to A Bets \$12, then the pay-off to A becomes \$4, so that A has been punished for his deviation. Likewise, if player A keeps his equilibrium strategy but player B changes to B Bets \$24, then the pay-off to A becomes \$8, worse for B than the equilibrium pay-off to A of \$6.

---

[6]The term "saddle point" is used because, at the center of a saddle, the saddle goes down in one direction (where the rider's legs are placed) and goes up in the perpendicular direction. Similarly, at a saddle point of a matrix, the values decrease in one direction and increase in the perpendicular direction.

The matrix of a game in normal form need not have any saddle point at all. This happens in the three matrices below.

$$\begin{pmatrix} 0 & 1 \\ 1 & 0 \end{pmatrix} \qquad \begin{pmatrix} 1 & -1 & 4 \\ 2 & 0 & 5 \\ 3 & 3 & 2 \end{pmatrix} \qquad \begin{pmatrix} 0 & 2 & 3 \\ 1 & 0 & 1 \end{pmatrix}$$

When the matrix of a game has no saddle point, then the game is inherently unstable, in the sense that, whichever strategies the players choose, one or both of the players would gain by switching to a different strategy. A game that has a saddle point is called a *strictly determined game*.

Example 2.3

Matching Pennies. Players A and B each have a penny. Without the opponent seeing it, each player places his penny on a table, with either the Head (H) or the Tail (T) showing. If the sides showing match (both Head or both Tail), then player A pays $1 to player B. If the sides do not match, player B pays $1 to player A. So, player A has two strategies: show a Tail (T_A) or show a Head (H_A), and, similarly, there are strategies (T_B) and (H_B) for player B. Then the pay-off table for player A is:

	T_B	H_B
T_A	−1	1
H_A	1	−1

Note that there is no saddle point. Now, if player A decides to play strategy (T_A) game after game, then, when player B observes what A is doing, B can play strategy (T_B), yielding a pay-off of −1, that is, A loses $1 to B. Likewise, if player A decides to play strategy (H_A) consistently, then, when B observes this, B can play strategy (H_B), yielding a pay-off of −1 again. Similarly, if B were to decide to consistently play one of his strategies, then A could choose a strategy to obtain a pay-off of $1 for A. Thus, neither player should stick to one particular strategy.

The situation in Example 2.3 is quite typical of what happens in games without saddle points. If one player consistently plays one strategy, the other player generally can adjust his strategy in order to increase his own pay-off. Therefore, it appears advisable for each player X not to consistently play one strategy, and instead to vary the strategies played so that opponent Y cannot guess what strategy X will choose to play next. The question is how best to do this. We shall return later to this rather complex problem and see that optimal solutions can be found for both players.

Example 2.4

When there is a saddle point in a game matrix, it can happen that there are
several such points. This occurs, for example, in the following matrices.

$$
\begin{pmatrix} 2 & 3 & 2 \\ 1 & 0 & 0 \\ 0 & 4 & -1 \end{pmatrix}
\qquad
\begin{pmatrix} 4 & 1 & 2 & 1 \\ 2 & 0 & 1 & -1 \\ 2 & 1 & 3 & 1 \\ 0 & 0 & 5 & 0 \end{pmatrix}
$$

In the first matrix there are two saddle points in the first row. In the second
matrix there are four saddle points (in the first row and second and fourth
columns, and in the third row, second and fourth columns). Notice that the
saddle points in each matrix have the same entries, 2 in the first matrix and
1 in the second matrix. These are instances of the following general property.

Theorem 2.1

1. If a matrix has saddle points, then all the entries at these saddle points
    are equal.
2. If (A_i, B_j) and (A_r, B_s) are saddle points, so are (A_i, B_s) and (A_r, B_j).

Proof. If the saddle points are in the same row, then their entries are
equal because they are the minimum entries in that row. Likewise, if the
saddle points are in the same column, their entries are equal because they
are the maximum entries in that column. So, assume now that we have two
entries a_{ij} and a_{rs} at saddle points in different rows and columns. Hence, $i \neq r$
and $j \neq s$. Since a_{ij} is minimal in its row, $a_{ij} \leq a_{is}$. Since a_{rs} is maximal in its column,
$a_{is} \leq a_{rs}$. Therefore, $a_{ij} \leq a_{rs}$. Similarly, by symmetry, reversing the order of the
saddle points and comparing their entries with a_{rj}, we deduce $a_{rs} \leq a_{ij}$.

Hence, from $a_{ij} \leq a_{rs}$ and $a_{rs} \leq a_{ij}$, we get $a_{ij} = a_{rs}$. Thus, all saddle points
have equal entries. Since we have shown that $a_{ij} \leq a_{is} \leq a_{rs} = a_{ij}$ and,
therefore, that $a_{ij} = a_{is} = a_{rs}$, it follows that a_{is} is minimal in its row and maximal
in its column. Thus, (A_i, B_s) is a saddle point. Similarly, (A_r, B_j) is a saddle
point. Thus, if two saddle points are not in the same row or column, then
the other two vertices of the rectangle determined by the two given saddle
points also must be saddle points.   ∎

When the matrix of a game in normal form has saddle points, then the
common value v of the entries at all the saddle points will be called the
value of the game. The value v is the pay-off for player A when both players
play the equilibrium pair of strategies (A_i, B_j) corresponding to any saddle
point. (Of course, the pay-off for player B will be $-v$.) For example, in the
game of Example 2.1, the value of the game is 2 and, in the game of Example
2.2, the value is 6. The value v is the pay-off for player A that will result if

the players do not want to risk obtaining a less than optimal pay-off. A player should play an equilibrium strategy when he is sure that his opponent knows how to find saddle points and understands their significance. If that assumption about the opponent is not true, then the player might gamble on getting a better pay-off and psychological or other factors may enter into the calculations.

When a game in normal form with saddle points has value $v = 0$, then the game is said to be *fair* since neither player has an advantage when they both choose their strategies wisely. Later on, we will be able to generalize the notions of value and fairness to games without saddle points.

Constant-sum games

If the sum of the pay-offs in a game is always a constant K, then the game is called a *constant-sum game*. A two-person constant-sum game G with constant sum K for the pay-offs is essentially equivalent to a two-person zero-sum game, in the sense that any question concerning the one game can be reduced to the other. In fact, define a new game G* in which, for each outcome of the original game, the new pay-off for each player is the old pay-off minus $K/2$. Clearly, G* is a zero-sum game and it is easy to see that questions about G (for example, existence of saddle points) are reducible to corresponding questions about G*.

Example 2.5

Duopoly. Two corporations A and B share a market of $100 million per year for a certain product. Assume that A has two marketing strategies and B three and, since their competition is a constant-sum game, we consider the corresponding zero-sum game and let their pay-offs be the difference between their annual sales and $50 million. Assume that the resulting pay-off table for player A (measured in millions of dollars) is:

	B_1	B_2	B_3
A_1	−40	20	−10
A_2	−20	0	10

There is one saddle point (A_2, B_1) with value −20. So, playing their optimal strategies, A's share of the market will be $30 million (that is, $50 million −$20 million) and B's share will be $70 million (that is, $50 million +$20 million).

Exercise 2.2

For each of the following matrices of games in normal form, determine whether there are saddle points. If there are, find them and the value of the game.

(a)   The two matrices of Example 2.4.

(b) $\begin{pmatrix} -1 & 2 & 0 \\ 0 & 3 & 4 \\ -2 & 1 & 4 \end{pmatrix}$ (c) $\begin{pmatrix} 2 & 0 \\ -1 & 7 \end{pmatrix}$ (d) $\begin{pmatrix} 3 & 0 & 3 \\ 3 & 1 & 3 \end{pmatrix}$ (e) $\begin{pmatrix} 2 & 1 & 1 & 1 \\ 3 & 1 & 2 & 1 \end{pmatrix}$

Exercise 2.3

In a matrix of a game in normal form with saddle points, is it possible for an entry to be equal to the value v of the game and not be a saddle point?

Exercise 2.4

A and B each has $100,000 in cash and they are equal partners in a business worth $60,000. They wish to dissolve the partnership and they agree that each of them will make a sealed bid, between $0 and $100,000, for the business. The higher bidder X gets the business and pays the other partner Y what X bid. If there is a tie in the bidding, the business will be sold for $60,000, which will be divided evenly. What bids should A and B make? (Hint: Let the pay-off for A be the value of what he had after the bidding minus what he had before the bidding. For example, if A bids $20,000 and B bids $40,000, then A gives up his share of the business, worth $30,000 and gets $40,000 from B, so that A's pay-off is $10,000 and B's pay-off is −$10,000.)

Exercise 2.5

For each of the following matrices, find out what x must be for the matrix to be strictly determined and indicate the value v of the game in those cases.

(a) $\begin{pmatrix} 0 & 3 \\ x & 1 \end{pmatrix}$ (b) $\begin{pmatrix} 0 & 3 \\ 1 & 3x \end{pmatrix}$ (c) $\begin{pmatrix} 1 & 3 & 4 \\ 2x & 4 & 1 \\ 0 & 5x & 2 \end{pmatrix}$

(d) $\begin{pmatrix} 1-x & x \\ 0 & x+1 \end{pmatrix}$ (e) $\begin{pmatrix} x & 2 & -x \\ 0 & x & -1 \end{pmatrix}$

2.3   Maximin and minimax

There is another useful way of looking at the pay-off matrix of a game in normal form. Let a_{ij} denote the entry in the ith row and jth column. For A's ith strategy A_i, find the minimum entry in the ith row and designate it $\min_j a_{ij}$. By playing strategy A_i, A can be sure of winning at least that amount. To maximize what he can be sure of winning in the game, A should choose a strategy that yields the largest value of $\min_j a_{ij}$. Such a strategy (there is at least one such) is called a *maximin strategy* for A, and the amount that he can be sure of winning when he plays that strategy can be written as $\max_i \min_j a_{ij}$ or simply as *maximin*. Similarly, for B's *j*th strategy B_j, find the maximum entry in the *j*th column and designate it $\max_i a_{ij}$. By playing strategy B_j, B can be sure of not losing more than that amount. To minimize what he might lose in the game, B should choose a strategy that yields the smallest value of $\max_i a_{ij}$. Such a strategy (there is at least one such) is called a *minimax strategy* for B, and the amount that is the most that he can lose when playing that strategy can be written as $\min_j \max_i a_{ij}$ or simply as *minimax*. By playing their maximin and minimax strategies, respectively, A can 'lock in' a certain minimum gain and B can 'lock in' a certain maximum loss.

Example 2.6

1. In the matrix

$$\begin{pmatrix} 1 & -1 & 4 \\ 2 & 0 & 5 \\ 3 & 3 & 2 \end{pmatrix}$$

the row minima are −1, 0, 2, so that the maximin is 2, attained in the third row. The column maxima are 3, 3, 5, so that the minimax is 3, attained in the first and second columns.

2. In the matrix

$$\begin{pmatrix} 4 & 1 & 2 & 1 \\ 2 & 0 & 1 & -1 \\ 2 & 1 & 3 & 1 \\ 0 & 0 & 5 & 0 \end{pmatrix}$$

the row minima are 1, −1, 1, 0, so that the maximin is 1, attained in the first and third rows. The column maxima are 4, 1, 5, 1, so that the minimax is 1, attained in the second and fourth columns.

There is a simple inequality between the maximin and the minimax.

Theorem 2.2

For any matrix, maximin ≤ minimax.

Proof. It is clear that every row minimum is less than or equal to every column maximum. (In fact, if u is the minimum in the ith row and w is the maximum in the jth column, and if c_{ij} is the entry in the ith row and jth column, then $u \le c_{ij} \le w$.) Hence, every row minimum is less than or equal to the minimax and, therefore, maximin $\le$ minimax.   ∎

The matrices in Example 2.6 show that, in Theorem 2.2, the strict inequality may or may not hold. Whether or not it holds is very important.

Theorem 2.3

For any matrix, maximin = minimax if and only if the matrix has a saddle point.

Proof. Let u = maximin and v = minimax. First assume that $u = v$. Let A_i be a row in which u is the minimum, and let B_j be a column in which v is the maximum. Let a_{ij} be the entry in the ith row and jth column. Then $u \le a_{ij}$ (since u is the minimum in the ith row) and $a_{ij} \le v$ (since v is the maximum in the jth column). But $u = v$. Therefore, $u = a_{ij} = v$. So, a_{ij} is the minimum in its row and the maximum in its column, that is, a_{ij} is a saddle point. Now, for the converse, assume that the matrix has a saddle point, and let the entry at a saddle point be a_{ij}, in the ith row and jth column. a_{ij} is the minimum in its row. Consider any other row, say, the kth row. Then the entry a_{kj} in the kth row and jth column satisfies $a_{kj} \le a_{ij}$, since the saddle point a_{ij} is the maximum in its column. But a_{kj} is at least as large as the minimum in the kth row. Hence, the minimum in the kth row is less than or equal to a_{ij}. Therefore, a_{ij} is the maximin. Now consider any column, say, the sth column. a_{ij} is the maximum in its column. The entry a_{is} in the ith row and sth column satisfies $a_{ij} \le a_{is}$ since the saddle point a_{ij} is the minimum in its row. But a_{is} is less than or equal to the maximum in its column. Hence, a_{ij} is less than or equal to the maximum in the sth column. Therefore, a_{ij} is the minimax. So, maximin $= a_{ij} =$ minimax. ∎

Example 2.7

1.

				row minima	maximin
1	−1	0	2	−1	
3	7	−2	4	−2	0
5	0	1	1	0	

column maxima 5   7   1   4
minimax   1

So, maximin < minimax. Hence, there is no saddle point, and that is, in fact, true.

2.

				row minima	maximin
1	−1	0	2	−1	
3	7	−2	4	−2	1
5	1	1	1	1	

column maxima 5   7   1   4
minimax   1

So, maximin = minimax. Hence, there is a saddle point, and, in fact, the entry 1 in the third row and third column is a saddle point. Note that the maximin is attained in the third row and the minimax is attained in the third column.

In general, a maximin strategy for player A and a minimax strategy for player B are the safest strategies. However, when the matrix of a game has no saddle point and the players play those maximin and minimax strategies, they will be led to an endless cycle of changes in their strategies. For example, consider the following table of a game with strategies A_1, A_2, A_3 for player A and strategies B_1, B_2, B_3 for player B:

	B_1	B_2	B_3
A_1	2	1	5
A_2	3	0	-1
A_3	0	2	2

Note that there is no saddle point. The maximin is 1, attained for strategy A_1. The minimax is 2, attained for strategy B_2. Now, assume that A and B start by playing their maximin and minimax strategies (A_1, B_2). The pay-off will be 1. Player A will not be satisfied with this result and would prefer the pay-off 2, which he can get by changing to strategy A_3. Therefore, when the game is played again, the strategies will be (A_3, B_2), yielding pay-off 2. Now, player B can get a better result by switching to strategy B_1. So, in the next game, the strategies will be (A_3, B_1) with pay-off 0. Similar reasoning shows that the succeeding games will be played with strategies (A_2, B_1), (A_2, B_3), (A_1, B_3), and then back to (A_1, B_2), where we originally started. So, the strategies will keep repeating in this way.

Exercise 2.6
For each of the following matrices, find the maximin and the minimax and determine whether there are saddle points. If there are, find them and the value of the game.

$$\text{(a)} \begin{pmatrix} 3 & 6 & 5 \\ 2 & 0 & 2 \\ 4 & 6 & 4 \end{pmatrix} \qquad \text{(b)} \begin{pmatrix} 2 & 7 \\ 3 & 1 \end{pmatrix} \qquad \text{(c)} \begin{pmatrix} 2 & 1 & 3 \\ 1 & 1 & 4 \end{pmatrix}$$

Throughout our discussion of how players A and B *should* play, we are assuming *rational behavior* on the part of the players. First of all, this means that the players know the pay-off matrix and have the logical and mathematical ability to follow the analysis that we are giving. Second, we assume that the players wish to maximize their gains and minimize their losses. Moreover, it is taken for granted that each player will attribute rational behavior to the other player. So, for example, player A has to assume that, if player A chooses strategy A_i, then B will select a strategy B_j that will minimize the pay-off $P(A_i,B_j)$ to player A. Thus, A can expect that the choice of strategy A_i will yield the minimum value in the ith row. (Of course, it might yield more if B makes a mistake; but A cannot depend on that happening.) Hence, since A wants to gain as much as possible, A will be impelled to play his maximin strategy, that is, the strategy A_i that maximizes $\min_j a_{ij}$. Similarly, B will wish to play his minimax strategy in order to minimize the amount that he can lose (and thus to maximize his gain).[7] From our earlier discussion, we know that, if the game matrix has a saddle point, then A's maximin strategy and B's minimax strategy yield an equilibrium pair.

Dominance

If each entry of one row X of the matrix of a game is greater than or equal to the entry in the same column of a second row Y, then row X is said to *dominate* row Y. Player A will never want to use the strategy of row Y since the strategy of row X always yields an equal or better pay-off. Therefore, the analysis of the game is simplified if row Y is eliminated. Similarly, if each entry of a column V is greater than or equal to the entry in the same row of a second column W, then column W is said to *dominate* column V. (Note that this is the opposite of dominance for rows.) Player B will never want to use the strategy of column V since the strategy of column W will always yield an equal or lesser gain for A (that is, an equal or greater gain for B). So, in such a case, the analysis of the game is simplified if column V is eliminated. Elimination of rows or columns by means of dominance relations often helps in the search for the best strategies for the players.

Example 2.8

1. Consider the 3×3 matrix on the left below.

$$\begin{pmatrix} -4 & 2 & -3 \\ -2 & -5 & 3 \\ -1 & 0 & 1 \end{pmatrix} \rightarrow \begin{pmatrix} -4 & 2 \\ -2 & -5 \\ -1 & 0 \end{pmatrix} \rightarrow \begin{pmatrix} -4 & 2 \\ -1 & 0 \end{pmatrix} \rightarrow \begin{pmatrix} -4 \\ -1 \end{pmatrix} \rightarrow (-1)$$

The first column dominates the third column. Elimination of the third column yields the next 3×2 matrix. Here, the third row dominates the second row. Elimination of the second row yields the next 2×2 matrix, in which the first column dominates the second column. Dropping the second column produces a 2×1 matrix, in which the second row dominates the first row. Crossing out the first row yields a 1×1 matrix which, in fact, is a saddle point of the original matrix.

[7]Remember that, according to our terminology, gains and losses can be negative. A negative gain is a loss and a negative loss is a gain. Hence, minimizing one's loss is equivalent to maximizing one's gain.

2. Note that the matrix

$$\begin{pmatrix} 3 & 5 & 8 \\ 2 & 1 & 9 \\ 1 & 6 & 0 \end{pmatrix}$$

has a saddle point 3 in the first row and first column but, in contra-distinction to the matrix of part (1), that saddle point cannot be found by using dominance relations, since, in this case, there are no domi-nance relations.

Exercise 2.7

1. Show that, if a 2×2 matrix has a saddle point, then that saddle point can be reached by using dominance relations.
2. Show that the result of part (1) holds for any $2 \times n$ or $m \times 2$ matrix.
3. Use dominance relations to simplify the following game matrices whenever possible, and then find saddle points (if any).

$$(i)\ \begin{pmatrix} 3 & 0 & -1 & 0 \\ 1 & 1 & -2 & 1 \\ 0 & 3 & 3 & 1 \\ 2 & 2 & 0 & 1 \end{pmatrix} \qquad (ii)\ \begin{pmatrix} 1 & 0 & 3 \\ 2 & 4 & 6 \\ 0 & 3 & -1 \end{pmatrix}$$

$$(iii)\ \begin{pmatrix} 1 & 5 & 3 \\ 2 & 4 & 1 \\ 0 & -1 & 1 \end{pmatrix} \qquad (iv)\ \begin{pmatrix} 1 & 2 & 3 \\ 0 & 3 & -1 \\ 0 & -1 & 4 \end{pmatrix}$$

Exercise 2.8

Two companies A and B want to build stores in one of three cities T, U, and V, whose locations and mutual distances are shown in the diagram below.

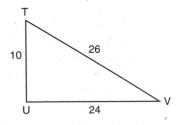

If they build in the same city, they split the total business of all three cities evenly. If they build in different cities, the company closer to a given city gets all of that city's business. All three cities yield the same amount of business. Where should they build? (Hint: Let c be the amount of business in each city. Each company has three strategies. Since this is a constant sum game (the sum is $3c$), change to a zero-sum game by letting the pay-off for company A be its total business minus $3c/2$. Write the pay-off matrix.)

Exercise 2.9

Players A and B alternately choose one of the numbers 1 or 2 until the sum of all the numbers chosen is at least 3. The last player to move wins from the other player the sum that has been reached. Construct the game tree, label all the positions, describe the players' strategies, and write the matrix for the game. Determine any dominance relations and find any saddle points.

Exercise 2.10

1. Show that a 2×2 matrix has no saddle point if and only if the entries on one diagonal are either both larger than or both smaller than the entries on the other diagonal.
2. Find a simple necessary and sufficient condition that the 2×2 matrix

$$\begin{pmatrix} 0 & c \\ d & 0 \end{pmatrix}$$

   has a saddle point.
3. Find 3×3 matrices with exactly n saddle points for $0 \leq n \leq 4$. What can be said about $n > 4$?

2.4 Mixed strategies

When a game matrix has no saddle point, it is inadvisable for either player to consistently play one of his strategies. We already have seen this happen in Example 2.3. Here are two more examples.

Example 2.9

Consider the following game matrix.

$$\begin{pmatrix} 2 & 0 \\ -1 & 1 \end{pmatrix}$$

The maximin is 0 and the minimax is 1, and there is no saddle point. If player A consistently plays either of his two strategies, then player B can choose a strategy that will yield to A at most the maximin 0. Likewise, if player B consistently plays one of his strategies, then player A can choose a strategy that will ensure A of a gain of at least the minimax 1. We shall see that, if A and B vary the choice of their strategies in certain random ways so that their opponent cannot guess how they will play, they will be assured of better pay-offs in the long run.

Example 2.10

Two-Finger Morra. Each player displays one or two fingers behind his back and, at the same time, guesses how many fingers his opponent shows. If both guesses are correct, the game is a draw. If only one player guesses correctly, that player receives the dollar sum of the fingers shown. Each player has four strategies: (1,1), (1,2), (2,1), (2,2), where (x,y) means that the player shows x and guesses y. Then the table of pay-offs for player A is:

	(1,1)	(1,2)	(2,1)	(2,2)
(1,1)	0	2	−3	0
(1,2)	−2	0	0	3
(2,1)	3	0	0	−4
(2,2)	0	−3	4	0

The maximin is −2 and the minimax is 2, and there is no saddle point. If player B plays a particular strategy, player A can choose a strategy that will assure A of a pay-off of at least the minimax 2. Likewise, if player A consistently plays a particular strategy, then player B can choose a strategy that will keep A's pay-off to at most the maximin −2, that is, it will assure B of a gain of at least 2. Thus, neither player should stick to one strategy, but should vary strategies in a random way that cannot be guessed by the opponent. We shall show later that both players can find an optimal way to do this.

It is apparent then that, when the game matrix has no saddle point, each player should vary the strategies he plays. Moreover, to prevent the opponent from using psychological or other factors to predict which strategy will be chosen, the strategies should be selected by a random device, such as coins, dice, a spinner, or even more sophisticated tools. For example, if a player has two strategies and wishes to employ them equally often but at random, then he can toss a fair coin to decide which strategy to use. If he has three strategies to choose among, he can use a spinner on a circle that is divided into three equal sectors, and so on.

Consider a game whose normal form is an $m \times n$ matrix. Thus, player A has m strategies $A_1, A_2, \ldots, A_m$ and player B has n strategies $B_1, B_2, \ldots, B_n$. By a *mixed strategy* for A we mean an m-tuple $(x_1, x_2, \ldots, x_m)$ of nonnegative real numbers $x_1, x_2, \ldots, x_m$ such that $x_1 + x_2 + \cdots + x_m = 1$. We shall say that player A plays the mixed strategy $(x_1, x_2, \ldots, x_m)$ if A plays strategy A_1 with probability x_1, strategy A_2 with probability $x_2, \ldots,$ and strategy A_m with probability x_m. Similarly, by a *mixed strategy* for player B we mean an n-tuple $(y_1, y_2, \ldots, y_n)$ of nonnegative real numbers $y_1, y_2, \ldots, y_n$ such that $y_1 + y_2 + \cdots + y_n = 1$, and we shall say that B plays the mixed strategy $(y_1, y_2, \ldots, y_n)$ if B plays strategy B_1 with probability y_1, strategy B_2 with probability $y_2, \ldots,$ and strategy B_n with probability y_n. Note that each player has infinitely many mixed strategies, except in the trivial case when a player has only one strategy.

Assume that the entry in the ith row and jth column of the $m \times n$ game matrix is denoted by c_{ij}. Thus, c_{ij} is the pay-off to player A when A plays strategy A_i and B plays strategy B_j. When A plays the mixed strategy

$(x_1, x_2, \ldots, x_m)$ and B plays the mixed strategy $(y_1, y_2, \ldots, y_n)$, then the probability that A plays strategy A_i and B plays strategy B_j is $x_i y_j$, since A and B choose their strategies independently.[8] So, the probability of getting the pay-off c_{ij} is $x_i y_j$. Therefore the expected pay-off $P(X,Y)$ to A when A plays strategy $X = (x_1, x_2, \ldots, x_m)$ and B plays strategy $Y = (y_1, y_2, \ldots, y_n)$ is the sum of all the terms $c_{ij} x_i y_j$, for all possible values of i and j. That sum can be written as

$$\sum_{1 \le i \le m, 1 \le j \le n} c_{ij} x_i y_j$$

Thus,

$$P(X, Y) = \sum_{1 \le i \le m, 1 \le j \le n} c_{ij} x_i y_j$$

Example 2.11
Consider the game of Example 2.9 with matrix

$$\begin{pmatrix} 2 & 0 \\ -1 & 1 \end{pmatrix}$$

Let us just see how to calculate two particular pay-offs. If player A plays mixed strategy $X_1 = (.5, .5)$ and player B plays mixed strategy $Y_1 = (.5, .5)$, then the pay-off $P(X_1, Y_1)$ to player A will be $2(.5)(.5) + 0(.5)(.5) + (-1)(.5)(.5) + 1(.5)(.5) = .5$. If player A plays mixed strategy $X_2 = (.4, .6)$ and player B plays mixed strategy $Y_2 = (.1, .9)$, the pay-off $P(X_2, Y_2)$ will be $2(.4)(.1) + 0(.4)(.9) + (-1)(.6)(.1) + 1(.6)(.9) = .56$.

Exercise 2.11
   1. In the game given by the matrix

$$\begin{pmatrix} -1 & 2 \\ 3 & 1 \end{pmatrix}$$

find the expected pay-off $P(X, Y)$ to A when A uses the mixed strategy $X = (1/3, 2/3)$ and B uses the mixed strategy $Y = (3/4, 1/4)$.
   2. If $X = (.2, .8)$ and $Y = (.5, 0, .5)$ are mixed strategies for A and B in the game given by the matrix

$$\begin{pmatrix} 2 & 1 & 3 \\ 1 & 1 & 4 \end{pmatrix}$$

compute $P(X, Y)$.

---

[8]When events E and F, with probabilities p and q, are independent, then the probability of the conjunction (E and F) is the product pq.

Now consider any game and a pair of mixed strategies (X^*, Y^*) for players A and B, respectively. We shall say that (X^*, Y^*) is an *equilibrium pair* if the following two conditions hold:

1. $P(X^*, Y^*) \leq P(X^*, Y)$ for any mixed strategy Y for player B;
2. $P(X^*, Y^*) \geq P(X, Y^*)$ for any mixed strategy X for player A.

This is analogous to the definition of an equilibrium pair corresponding to a saddle point of a game matrix. Condition (1) says that $P(X^*, Y^*)$ is a minimum in its 'row', where the 'row' consists of the numbers $P(X^*, Y)$ for all mixed strategies Y for player B. Similarly, condition (2) says that $P(X^*, Y^*)$ is a maximum in its 'column', where the 'column' consists of the numbers $P(X, Y^*)$ for all mixed strategies X for player A.

An equilibrium pair is *stable* in the sense that it is unwise for a player to deviate from his mixed strategy in an equilibrium pair if his opponent plays the opponent's mixed strategy of that pair. The reason for this is the same as that given for the stability of equilibrium pairs corresponding to saddle points.

In contradistinction to mixed strategies, the original strategies of the players are called *pure strategies*. Every pure strategy also can be thought of as a mixed strategy. For example, if player A has the pure strategies $A_1, A_2, \ldots,$ A_m, then the pure strategy A_1 can be represented by the mixed strategy $(1, 0, 0, \ldots, 0)$, the pure strategy A_2 can be represented by the mixed strategy $(0, 1, 0, \ldots, 0)$, and so on. Note that a pay-off $P(A_i, B_j)$ for pure strategies A_i and B_j turns out to be the same as the pay-off when the associated mixed strategies play against each other.

Recall that certain game matrices do not possess saddle points and equilibrium pairs. However, the following remarkable result holds:

Von Neumann's Theorem
Every game matrix has at least one equilibrium pair of *mixed strategies*.[9]

We shall return later (in Section 7 and Chapter 3) to this theorem and its proof via the theory of linear programming. In the meantime, we shall derive some important general facts about equilibrium pairs. We shall also

---

[9]This theorem is sometimes called the Minimax Theorem and was first proved in von Neumann [1928], using rather sophisticated results from topology. A proof using much simpler mathematical ideas, but still retaining a slight topological flavor, was given in Ville [1938], where the theorem was extended to continuous games (in which the players can have continuum-many strategies). The proof that appears in the 1944 book by von Neumann and Morgenstern essentially follows Ville's approach. The first purely algebraic proof was found by Loomis [1946].

John von Neumann was the founder of mathematical game theory. His 1928 paper was the first to apply mathematics to the study of games in a consistent and precise way. (The French mathematician Emile Borel was making some progress in the same direction, but did not know that von Neumann's Theorem was valid.) However, it was his 1944 book, written with the economist Oskar Morgenstern, that first brought game theory to the attention of mathematicians and economists. He was a mathematical genius of the highest order and also made fundamental contributions in physics and computer science, in both theory and practice. (See Halmos [1973].)

handle the special cases of 2×2 and $2 \times n$ matrix games before dealing with the general case. At a few places, we shall obtain some results that depend on von Neumann's Theorem, but we will not be guilty later of circular reasoning, since those results are not used in the proof of von Neumann's Theorem.

From now on, "equilibrium pair" will mean "equilibrium pair of mixed strategies", unless something is said to the contrary. If we wish to refer to equilibrium pairs of the original strategies, then we shall call them "equilibrium pairs of pure strategies".

Theorem 2.4

If (X_1, Y_1) and (X_2, Y_2) are equilibrium pairs of a game, then

$$P(X_1, Y_1) = P(X_2, Y_2).$$

Proof. $P(X_1, Y_1) \leq P(X_1, Y_2) \leq P(X_2, Y_2) \leq P(X_2, Y_1) \leq P(X_1, Y_1)$. Since $P(X_1, Y_1) \leq P(X_2, Y_2)$ and $P(X_2, Y_2) \leq P(X_1, Y_1)$, it follows that $P(X_1, Y_1) = P(X_2, Y_2)$. ∎

By virtue of Theorem 2.4, the pay-offs for all equilibrium pairs $P(X, Y)$ of a game are equal. That common value is called the *value v* of the game (for player A). Since, by von Neumann's Theorem, all games will turn out to have equilibrium pairs of mixed strategies, every game will have a value. As in the case of games with saddle points, a game is called *fair* if its value is 0. Since the value is the pay-off when the players play equilibrium pair strategies, and such strategies can be considered to be optimal, a fair game gives no advantage to either player in the long run.

Theorem 2.5

If (X_1, Y_1) and (X_2, Y_2) are equilibrium pairs of a game, then so is (X_1, Y_2).

Proof. Assume that (X_1, Y_1) and (X_2, Y_2) are equilibrium pairs. By Theorem 2.4, $P(X_1, Y_1) = P(X_2, Y_2)$ and, from the proof of that theorem, $P(X_1, Y_2) = P(X_1, Y_1)$. So, $P(X_1, Y_2) = P(X_1, Y_1) \leq P(X_1, Y)$ for any mixed strategy Y for B. Also, $P(X_1, Y_2) = P(X_2, Y_2) \geq P(X, Y_2)$ for any mixed strategy X for A. Thus, (X_1, Y_2) is an equilibrium pair.   ∎

A mixed strategy X for player A is called an *optimal strategy* for A if there is a mixed strategy Y for player B such that (X, Y) is an equilibrium pair. Likewise, a mixed strategy Y for player B is called an *optimal strategy* for B if there is a mixed strategy X for player A such that (X, Y) is an equilibrium pair. By Theorem 2.5, *any* optimal strategy X for A combines with *any* optimal strategy Y for B to give an equilibrium pair (X, Y).

Exercise 2.12

Let v be the value of a matrix game. Prove: If X^* is an optimal strategy for player A, then $v \leq P(X^*, Y)$ for any mixed strategy Y for player B, and, if Y^* is an optimal strategy for player B, then $v \geq P(X, Y^*)$ for any mixed strategy X for player A.

Given a mixed strategy $X = (x_1, \ldots, x_m)$ for player A in a game, define

$$u(X) = \min_{all\ Y} P(X, Y)$$

where the minimum is taken over all mixed strategies Y for player B. Thus, *u(X) is the minimum pay-off that player A can be guaranteed using strategy X.* To see that $u(X)$ exists, first note that, if c_{ij} is the entry in the *i*th row and *j*th column of the game matrix and an arbitrary strategy Y for player B is represented by $(y_1, \ldots, y_n)$, then

$$u(X) = \min_{all\ Y} \left(\sum_{1 \leq i \leq m, 1 \leq j \leq n} c_{ij} x_i y_j \right)$$

The set of all mixed strategies Y for player B is the set of all *n*-tuples $(y_1, \ldots, y_n)$ of nonnegative real numbers satisfying $y_1 + \cdots + y_n = 1$. This set is a closed, bounded subset of R^n and, for fixed $x_1, \ldots, x_m$, the function $\sum_{1 \leq i \leq m, 1 \leq j \leq n} c_{ij} x_i y_j$ is continuous on that set. Hence, that function has a minimum value $u(X)$ on that set.[10]

Now, for any mixed strategy $Y = (y_1, \ldots, y_n)$ for player B, define $w(Y) = \max_{all\ X} P(X, Y)$ where the maximum is taken over all mixed strategies X for player A. Thus, $w(Y)$ is the maximum pay-off that player A can hope to get when player B uses strategy Y; so, B knows that he will lose no more than $w(Y)$ when he uses strategy Y. To see that $w(Y)$ exists, first note that, if c_{ij} is the entry in the *i*th row and *j*th column of the game matrix and an arbitrary strategy X for player A is represented by $(x_1, \ldots, x_m)$, then

$$w(Y) = \max_{all\ X} \left(\sum_{1 \leq i \leq m, 1 \leq j \leq n} c_{ij} x_i y_j \right)$$

The set of all mixed strategies X for player A is the set of all *m*-tuples $(x_1, \ldots, x_m)$ of nonnegative real numbers satisfying $x_1 + \cdots + x_m = 1$. This set is a closed, bounded subset of R^m and, for fixed $y_1, \ldots, y_n$, the function $\sum_{1 \leq i \leq m, 1 \leq j \leq n} c_{ij} x_i y_j$ is continuous on that set. Hence, that function has a maximum value $w(Y)$ on that set.

---

[10]In general, a continuous function on a closed, bounded subset of R^n has a minimum and a maximum value on that set. A subset B of R^n is said to be *bounded* if there is a number M such that $|X| \leq M$ for every point X in B. B is said to be *closed* if all limit points X of B (if any) belong to B. (Here, X is a *limit point* of B if there are points of B arbitrarily close to X, that is, for any positive real number ε, there is a point Y in B such that $|X - Y| < \varepsilon$.) Examples of closed bounded sets in R^2 are disks, boxes, and, generally, any set consisting of the boundary and interior of a simple closed curve.

$u(X)$ is an analogue of the notion of the minimum along a row of the original game matrix, except that here the 'row' consists of all the pay-offs $P(X,Y)$, where Y varies over all mixed strategies for player B. $w(Y)$ is an analogue of the notion of the maximum in a column of the original game matrix, except that here the 'column' consists of all the pay-offs $P(X,Y)$, where X varies over all mixed strategies for player A. The following result is an analogue of the fact that the minimum along a row is no greater than the maximum in a column.

Theorem 2.6
For any X and Y, $u(X) \le w(Y)$.

Proof. $u(X) \le P(X,Y) \le w(Y)$.   ∎

The central connection between equilibrium pairs and the $u(X)$ and $w(Y)$ functions is given by the following theorem.

Theorem 2.7
For any X^* and Y^*, (X^*,Y^*) is an equilibrium pair if and only if $u(X^*) = w(Y^*)$. Moreover, for any equilibrium pair (X^*,Y^*), $u(X^*) = w(Y^*) = P(X^*,Y^*) = v$, where v is the value of the game.

Proof
1. Assume (X^*,Y^*) is an equilibrium pair. Then $P(X^*,Y^*)$ is the maximum of $P(X,Y^*)$ with respect to all X, that is, $P(X^*,Y^*) = w(Y^*)$. Similarly, $P(X^*,Y^*)$ is the minimum of $P(X^*,Y)$ with respect to all Y, that is, $P(X^*,Y^*) = u(X^*)$. Hence, $u(X^*) = P(X^*, Y^*) = w(Y^*)$.
2. Assume $u(X^*) = w(Y^*)$. Then, for all Y, we have $P(X^*,Y^*) \le w(Y^*) = u(X^*) \le P(X^*,Y)$ and, for all X, $P(X^*,Y^*) \ge u(X^*) = w(Y^*) \ge P(X, Y^*)$. So, (X^*,Y^*) is an equilibrium pair.   ∎

Exercise 2.13
Let X be a mixed strategy for player A, let Y be a mixed strategy for player B, and let v be the value of the game. Show that X is optimal for A if and only if $u(X) = v$, and Y is optimal for B if and only if $w(Y) = v$.

Theorem 2.8
Assume (X^*,Y^*) is an equilibrium pair. Then:

1. $u(X^*) \ge u(X)$ for all X,
2. $w(Y^*) \le w(Y)$ for all Y.

Thus, by (1), $u(X^*) = \max_X u(X) = \max_X (\min_Y P(X, Y))$ is the maximin and, by (2), $w(Y^*) = \min_Y w(Y) = \min_Y (\max_X P(X, Y))$ is the minimax, and, by Theorem 2.7, the maximin = the minimax.

Proof.   By Theorems 2.6 and 2.7, $u(X) \leq w(Y^*) = u(X^*)$ and $w(Y) \geq u(X^*) = w(Y^*)$.   ■

Corollary 2.9
For any mixed strategy X for player A and any mixed strategy Y for player B, $u(X) \leq v \leq w(Y)$, where v is the value of the game.

Proof.   Use Theorems 2.7 and 2.8.   ■

Consider any game in which player A has strategies $A_1, \ldots, A_m$ and player B has strategies $B_1, \ldots, B_n$, and let c_{ij} be the entry in the ith row and jth column of the game matrix. For any mixed strategy $X = (x_1, \ldots, x_m)$ for player A, we know that

$$u(X) = \min_{\text{all } Y} \left(\sum_{1 \leq i \leq m, 1 \leq j \leq n} c_{ij} x_i y_j \right)$$

where $Y = (y_1, \ldots, y_n)$ represents an arbitrary mixed strategy for player B. Now the expression on the right can be rewritten as $\min_{\text{all } Y} (\sum_{j=1}^{n} y_j (\sum_{i=1}^{m} c_{ij} x_i))$. Clearly, the minimum is attained when we let $y_j = 1$ for that j which yields the smallest value for $(\sum_{i=1}^{m} c_{ij} x_i)$ and 0 for the other coordinates. Thus, $u(X) = \min_{1 \leq j \leq n} \sum_{i=1}^{m} c_{ij} x_i$. But, $\sum_{i=1}^{m} c_{ij} x_i$ represents the pay-off $P(X, B_j)$, since the mixed strategy corresponding to the pure strategy B_j has 1 as its jth coordinate and 0 as all the other coordinates. So, we obtain the following important results.

Theorem 2.10

$$u(X) = \min_{1 \leq j \leq n} P(X, B_j) = \text{minimum } (P(X, B_1), \ldots, P(X, B_n))$$

By a similar argument, we get:

Theorem 2.11

$$w(Y) = \max_{1 \leq i \leq m} P(A_i, Y) = \text{maximum } (P(A_1, Y), \ldots, P(A_m, Y))$$

Example 2.12
Let $X = (.5, .5, 0)$ and $Y = (.3, .5, .2)$ be mixed strategies for players A and B in the game given by the matrix

$$\begin{pmatrix} 3 & 1 & 2 \\ 4 & 1 & 1 \\ 2 & 2 & 0 \end{pmatrix}$$

Let A_1, A_2, A_3 be A's strategies, and let B_1, B_2, B_3 be B's strategies. Then computation yields $P(X,B_1) = 3.5$, $P(X,B_2) = 1$, $P(X,B_3) = 1.5$. Hence, by Theorem 2.10, $u(X) = $ min $(3.5, 1, 1.5) = 1$. Likewise, $P(A_1,Y) = 1.8$, $P(A_2,Y) = 1.9$, $P(A_3,Y) = 1.6$, and, therefore, by Theorem 2.11, $w(Y) = $ max $(1.8, 1.9, 1.6) = 1.9$. Since $u(X) \neq w(Y)$, (X,Y) is not an equilibrium pair.

Example 2.13

Recall the pay-off table for Matching Pennies (Example 2.3):

	T_B	H_B
T_A	-1	1
H_A	1	-1

The matrix has no saddle point. Let $X = (^1/_2, {}^1/_2)$ be the mixed strategy for player A in which Tails and Heads are each chosen with probability $^1/_2$. Let $Y = (^1/_2, {}^1/_2)$ be the mixed strategy for player B in which Tails and Heads each has probability $^1/_2$. Now, $P(X, T_B) = {}^1/_2(-1) + {}^1/_2(1) = 0$ and $P(X, H_B) = {}^1/_2(1) + {}^1/_2(-1) = 0$. So, $u(X) = $ min $(P(X, T_B), P(X, H_B)) = $ min $(0,0) = 0$. Similarly, $P(T_A, Y) = {}^1/_2(-1) + {}^1/_2(1) = 0$ and $P(H_A,Y) = {}^1/_2(1) + {}^1/_2(-1) = 0$, and $w(Y) = $ max $(P(T_A,Y)), P(H_A,Y)) = $ max $(0,0) = 0$. Hence, $u(X) = w(Y) = 0$. So, (X, Y) is an equilibrium pair and the value v of the game is 0. Since $v = 0$, the game is fair. The players should choose their strategies by flipping a fair coin.

Example 2.14

Scissors, Stone, Paper. This is a children's game in which each player calls out "Scissors," "Stone," or "Paper." The rule is that Scissors cuts Paper, Paper covers Stone, and Stone blunts Scissors,[11] and the pay-off table can be set up as follows:

	Sc_B	St_B	P_B
Sc_A	0	-1	1
St_A	1	0	-1
P_A	-1	1	0

Sc_A is the strategy in which player A calls out "Scissors," and similarly for the rest of the notation. There are no saddle points. Let $X = (^1/_3, {}^1/_3, {}^1/_3)$ and $Y = (^1/_3, {}^1/_3, {}^1/_3)$ be mixed strategies for players A and B, respectively. Then, $P(X,Sc_B) = {}^1/_3(0) + {}^1/_3(1) + {}^1/_3(-1) = 0$ and, similarly, $P(X,St_B) = P(X,P_B) = 0$. So $u(X) = $ min$(P(X,Sc_B), P(X,St_B), P(X,P_B)) = 0$. Likewise, $P(Sc_A,Y) = P(St_A,Y) = P(P_A,Y) = 0$ and $w(Y) = 0$. Thus, $u(X) = w(Y) = 0$. Hence, (X,Y) is an equilibrium pair, the value $v = 0$, and the game is fair. The players should play their three strategies with equal frequencies of $^1/_3$.

---

[11] A Chinese version of the same game reads: Man eats rooster, rooster eats worm, worm eats man.

Example 2.15

Simple Poker. This is a card game that shows in a very simple way some of the factors, such as bluffing, that are involved in playing Poker. In this game, one card is dealt at random to player A from a normal deck of cards. Player A observes the card, which B does not see. Then either A 'folds' (that is, gives up) and pays $1 to player B, or A 'bets' $3. If A bets, then B either 'folds' and pays $1 to A, or B can 'call'. If B calls, A wins $3 from B if A's card is red or loses $3 to B if A's card is black. In this game, A has the following four strategies: *(Bet, Bet)* Bet no matter what color the card is; *(Bet, F)* Bet if the card is red and fold if it is black; *(F, Bet)* Fold if the card is red and Bet if it is black; *(F, F)* Fold no matter what color the card is. On the other hand, B has only two strategies: *(F_B)* Fold if A bets; *(C_B)* Call if A bets. Here is the pay-off table (for A's winnings, as usual).

	F_B	C_B
(Bet,Bet)	1	0
(Bet, F)	0	1
(F, Bet)	0	−2
(F, F)	−1	−1

Let us look at the computation of some of the pay-offs. Consider the first row, first column. Since A bets in all cases, B will always fold and A will win $1. Consider the first row, second column. Player A always bets. A red card appears with probability $1/2$, and then player A will bet, B will call, and A wins $3. Thus, A wins $3 with probability $1/2$. A black card appears with probability $1/2$, and then A will bet, B will call, and A loses $3. So, A's gain is −$3 with probability $1/2$. Hence, the expected pay-off for A will be $1/2$ (3) + $1/2$ (−3) = 0. Now consider one more entry, the third row, second column, that is, (F, Bet) vs. C_B. If a red card appears (this happens with probability $1/2$), then A folds and A loses $1 to B. If a black card appears (this happens with probability $1/2$), A will bet, B will call, and A will lose $3. Hence, A's expected pay-off is $1/2(−1) + 1/2(−3) = −2$. The calculation of the other five entries is left to the reader. There are no saddle points. The third and fourth rows are dominated by the second row (and by the first row); so we can be sure that player A will never play strategies (F, Bet) and (F, F). Thus, the table reduces to:

	F_B	C_B
(Bet,Bet)	1	0
(Bet, F)	0	1

The symmetry of this matrix makes it plausible that players A and B will play their two strategies with equal probability. Hence, we test $X = (1/2, 1/2, 0, 0)$ and $Y = (1/2, 1/2)$ to see whether (X, Y) is an equilibrium pair. Now, $P(X, F_B) = 1/2$ and $P(X, C_B) = 1/2$. So, $u(X) = \min (1/2, 1/2) = 1/2$. Similarly, $P((Bet, Bet), Y) = 1/2$ and $P((Bet, F), Y) = 1/2$. So, $w(Y) = \max(1/2, 1/2, 0, 0) = 1/2$. Thus, $u(X) = 1/2 = w(Y)$. Therefore, (X, Y) is an equilibrium pair and the value v is $1/2$. Note that the game is not fair; it is biased in favor of A, since he has a long-run expected gain of half a dollar per game. Observe also that A's strategy X involves occasional bluffing, since the strategy *(Bet, Bet)* sometimes requires

A to bet even when A has a losing card. Player A must use (*Bet, Bet*) sometimes, since, if A always played strategy (*Bet, F*), then B could use strategy
C_B and have an expected gain of 1.

In Examples 2.13 – 2.15, we depended upon our intuitions about sym-
metries in the game matrices to guess the correct equilibrium pairs. We shall
have to look for techniques that yield equilibrium pairs and do not depend
upon our intuition or guesswork.

Exercise 2.14
Assume that (*X*, Y**) is an equilibrium strategy for a matrix game.

1. If all entries of the given matrix are increased by a constant K, show
    that (*X*,Y**) is an equilibrium strategy for the new game and the
    value of the game is increased by K. (Hint: Show that all $u(X)$ and
    $w(Y)$ are increased by K.)
2. If all entries of the matrix are multiplied by a positive constant C,
    show that (*X*,Y**) is an equilibrium strategy for the new game and
    the value of the game is multiplied by C. (Hint: Show that all $u(X)$
    and $w(Y)$ are multiplied by C.)

Theorem 2.12
Let player A have strategies $A_1,..., A_m$ and player B have strategies $B_1,..., B_n$,
let $X^* = (x_1,..., x_m)$ and $Y^* = (y_1,..., y_n)$ form an equilibrium pair and let $v =
P(X^*,Y^*)$ be the value of the game. If $y_k > 0$, then $P(X^*,B_k) = v$, and, similarly,
if $x_r > 0$, then $P(A_r,Y^*) = v$.

> *Proof.*   Let $C = (c_{ij})$ be the game matrix. We know, by Exercise 2.12, that
> $v \le P(X^*,B_j) = \sum_{i=1}^{m} c_{ij}x_i$ for $1 \le j \le n$. For the sake of contradiction, assume $v <
> P(X^*,B_k) = \sum_{i=1}^{m} c_{ik}x_i$. Then $v = P(X^*, Y^*) = (\sum_{j=1}^{m}y_j \, (\sum_{i=1}^{m}c_{ij}x_i)) > \sum_{j=1}^{m}y_jv = v$,
> which is a contradiction.   ∎

Exercise 2.15
1. If X_1 and X_2 are in R^m, then the *line segment* between X_1 and X_2 consists
    of all $tX_1 + (1 - t)X_2$ for $0 \le t \le 1$.
2. If X_1 and X_2 are optimal strategies for player A, then, for $0 \le t \le 1$,
    $tX_1 + (1 - t)X_2$ is a mixed strategy that is optimal for A. (Hint: Let $X^*
    = tX_1 + (1 - t)X_2$ and let Y^* be optimal for player B. Then $P(X^*, Y^*)
    = tP(X_1, Y^*) + (1 - t)P(X_2, Y^*) = tv + (1 - t)v = v$. Since Y^* is optimal,
    $P(X, Y^*) \le v = P(X^*, Y^*)$ for any X. Also, for any Y, $P(X^*, Y) = tP(X_1,
    Y)+ (1 - t)P(X_2, Y) \ge tv + (1 - t)v = v = P(X^*, Y^*)$, by Exercise 2.12.
    Thus, (*X*, Y**) is an equilibrium pair.)

By the *support* of a mixed strategy we mean the set of pure strategies
that occur with positive probability in the given mixed strategy. For example,
if $A_1,..., A_6$ are the strategies for player A in a given game, then the support
of the mixed strategy (0,1/3,1/6,1/6,0,1/3) is {A_2, A_3, A_4, A_6}. Theorem 2.12 tells

us that, if A_r belongs to the support of an optimal strategy for player A, then, for any optimal strategy Y^* for player B, $P(A_r, Y^*) = v$, where v is the value of the game. Similarly, if B_k belongs to the support of an optimal strategy for B, then, for any optimal strategy X^* for A, $P(X^*, B_k) = v$.

Exercise 2.16

The ith row of a matrix is said to be *strictly dominated* by the jth row if the jth row dominates the ith row and is different from the ith row. Show that, if the ith row is strictly dominated by another row, then the ith strategy A_i for player A is not in the support of any optimal strategy for A. (A similar result holds for columns and player B.)

Exercise 2.17

Let player A have strategies $A_1, \ldots, A_m$ and player B have strategies $B_1, \ldots, B_n$ and assume that the game matrix $C = (c_{ij})$ has a saddle point at the entry c_{rs}. Show that (A_r, B_s) is an equilibrium pair (with respect to arbitrary mixed strategies). (Hint: Using the fact that c_{rs} is the minimum in its row and the maximum in its column, apply Theorems 2.10 and 2.11 to show that $u(A_r) = c_{rs}$ and $w(B_s) = c_{rs}$.)

There is a large class of games, called *symmetric games*, in which the rules of the game treat players A and B in exactly the same way. If player A has n strategies $A_1, \ldots, A_n$, then player B will have n corresponding strategies $B_1, \ldots, B_n$ and the pay-off $P(A_i, B_j)$ for player A is the same as the pay-off for B when B plays strategy B_i against A's strategy A_j. Since the latter is $-P(A_j, B_i)$, we have $P(A_i, B_j) = -P(A_j, B_i)$. So, the game matrix $C = (c_{ij})$ satisfies the conditions $c_{ij} = -c_{ji}$ for $1 \leq i, j \leq n$. Such a matrix is said to be *skew-symmetric*. Note that, when $i = j$, the condition $c_{ij} = -c_{ji}$ becomes $c_{ii} = -c_{ii}$, which entails $c_{ii} = 0$. Hence, the principal diagonal of a skew-symmetric matrix consists of 0's. Examples of symmetric games are two-finger Morra (Example 2.10) and scissors-stone-paper (Example 2.14). Since neither player in a symmetric game has an advantage, the following result should not be surprising.

Theorem 2.13

The value v of a symmetric game must be zero.

Proof. $u(X) = \min\limits_{1 \leq j \leq n} P(X, B_j) = \min\limits_{1 \leq j \leq n} \sum_{i=1}^{n} c_{ij} x_i = \min\limits_{1 \leq j \leq n} - \sum_{i=1}^{n} c_{ji} x_i =$ $-\max\limits_{1 \leq j \leq n} \sum_{i=1}^{n} c_{ji} x_i = -w(X)$. Therefore, $u(X)$ and $w(X)$ have opposite signs. Since $u(X) \leq w(Y)$ for all X and Y, $u(X) \leq w(X)$. So, $u(X) \leq 0 \leq w(X)$. Let (X, Y) be an equilibrium pair. $u(X) = w(Y)$. But $u(X) \leq 0$ and $0 \leq w(Y)$. Hence, $u(X) = 0 = w(Y)$, and, since $v = u(X)$, $v = 0$. ∎

Example 2.16

Consider the symmetric matrix game defined by the skew-symmetric matrix:

$$\begin{pmatrix} 0 & 2 & -3 \\ -2 & 0 & 4 \\ 3 & -4 & 0 \end{pmatrix}$$

Let us find an equilibrium pair (X, Y). Let $X = (x_1, x_2, x_3)$. By Theorem 2.13, the value $v = 0$. Let B_1, B_2, B_3 be player B's strategies. By Exercise 2.12, $0 \leq P(X, B_1) = -2x_2 + 3x_3$, $0 \leq P(X, B_2) = 2x_1 - 4x_3$, and $0 \leq P(X, B_3) = -3x_1 + 4x_2$. So, $2x_2 \leq 3x_3$, $4x_3 \leq 2x_1$, and $3x_1 \leq 4x_2$, whence $^2/_3\, x_2 \leq x_3 \leq {}^1/_2\, x_1 \leq {}^2/_3 x_2$. Thus, $^2/_3\, x_2 = x_3 = {}^1/_2\, x_1$. Since $x_1 + x_2 + x_3 = 1$, we get $1 = x_1 + {}^3/_4\, x_1 + {}^1/_2\, x_1 = (9/4)x_1$. Then $x_1 = 4/9$, $x_2 = 3/9$, $x_3 = 2/9$. So, $X = (4/9,3/9,2/9)$ and, by symmetry, $Y = (4/9,3/9,2/9)$.

Example 2.17

Let us look at the symmetric matrix game of two-finger Morra in Example 2.10. Let $X = (x_1, x_2, x_3, x_4)$ be an optimal strategy for player A. Let B_1, B_2, B_3, B_4 be player B's strategies. Since the game is symmetric, $v = 0$, and, by Exercise 2.12, $v \leq P(X, B_j)$ for $1 \leq j \leq 4$, that is, $0 \leq -2x_2 + 3x_3$, $0 \leq 2x_1 - 3x_4$, $0 \leq -3x_1 + 4x_4$, $0 \leq 3x_2 - 4x_3$. Multiplying the second and third inequalities by 3 and 2, respectively, and then adding, we get $0 \leq -x_4$. Since $x_4 \geq 0$, it follows that $x_4 = 0$. So, the third inequality yields $0 \leq -3x_1$, whence $x_1 = 0$. The first and fourth inequalities yield $2x_2 \leq 3x_3$ and $4x_3 \leq 3x_2$. Since $x_1 + x_2 + x_3 + x_4 = 1$ and $x_1 = x_4 = 0$, we have $x_3 = 1 - x_2$ and then the inequalities $2x_2 \leq 3x_3$ and $4x_3 \leq 3x_2$ imply $4/7 \leq x_2 \leq 3/5$. So, the optimal strategies for player A are all X of the form $(0, x, 1 - x, 0)$, where $4/7 \leq x \leq 3/5$. By symmetry, player B has optimal strategies of the same form. (Notice that the optimal strategies never use (1, 1) or (2, 2), that is, each player should never guess that the opponent has chosen the same number that he himself has chosen.)

Exercise 2.18

Find all equilibrium pairs for the symmetric games with the following matrices:

(a) $\begin{pmatrix} 0 & 1 & -2 \\ -1 & 0 & 2 \\ 2 & -2 & 0 \end{pmatrix}$ (b) $\begin{pmatrix} 0 & 3 & -1 \\ -3 & 0 & 4 \\ 1 & -4 & 0 \end{pmatrix}$ (c) $\begin{pmatrix} 0 & -1 & 2 \\ 1 & 0 & -3 \\ -2 & 3 & 0 \end{pmatrix}$

(d) $\begin{pmatrix} 0 & 4 & -1 & 2 \\ -4 & 0 & -3 & 0 \\ 1 & 3 & 0 & -1 \\ -2 & 0 & 1 & 0 \end{pmatrix}$ (e) $\begin{pmatrix} 0 & -1 & -2 & 3 \\ 1 & 0 & -1 & 4 \\ 2 & 1 & 0 & -1 \\ -3 & -4 & 1 & 0 \end{pmatrix}$

Exercise 2.19

Show that every symmetric 2×2 matrix game has a saddle point but that there are symmetric 3×3 matrix games without saddle points.

Exercise 2.20

1. Find an equilibrium pair for the symmetric 3×3 matrix game

$$\begin{pmatrix} 0 & 1 & -4 \\ -1 & 0 & 2 \\ 4 & -2 & 0 \end{pmatrix}$$

2. Let

$$\begin{pmatrix} 0 & a & b \\ -a & 0 & c \\ -b & -c & 0 \end{pmatrix}$$

be a symmetric 3×3 matrix game without a saddle point.
(i) Show that a, b, c are nonzero, a and c have the same sign, and a and b have opposite sign.
(ii) Letting $M = |a| + |b| + |c|$, show that an equilibrium pair is given by $X = Y = (|c|/M, |b|/M, |a|/M)$.

Equilibrium pairs of games determined by higher-order matrices can be found, as in Examples 2.16 and 2.17, by calculations based on Exercise 2.12, or by using Theorem 2.12 and solving the resulting system of equations. Sometimes such computations can be quite tedious. More systematic procedures using linear programming techniques will be introduced later on.

Example 2.17
Let us find an equilibrium pair $X = (x_1, x_2, x_3)$, $Y = (y_1, y_2, y_3)$ for the 3×3 matrix

$$\begin{pmatrix} 0 & -1 & 2 \\ 3 & 1 & 0 \\ -2 & 2 & 1 \end{pmatrix}$$

Assume first all $y_i > 0$. Then, by Theorem 2.12,

$$P(X, B_1) = 3x_2 - 2x_3 = v$$

$$P(X, B_2) = -x_1 + x_2 + 2x_3 = v$$

$$P(X, B_3) = 2x_1 + x_3 = v \quad \text{and} \quad x_1 + x_2 + x_3 = 1$$

Solving, we obtain $x_1 = 6/24$, $x_2 = 11/24$, $x_3 = 7/24$, $v = 19/24$. Since each $x_i > 0$, we can use Theorem 2.12 in the same way to obtain $y_1 = 4/24$, $y_2 = 7/24$, $y_3 = 13/24$. (It can be shown by methods developed below that there is no other equilibrium pair.)

Exercise 2.21
Find equilibrium pairs and the value v for the following matrices.

$$\text{(a)} \begin{pmatrix} 1 & 1 & 3 \\ 2 & 1 & 2 \\ 3 & 4 & 1 \end{pmatrix} \quad \text{(b)} \begin{pmatrix} 2 & 1 & -1 \\ 0 & 2 & 3 \\ 1 & 0 & 1 \end{pmatrix} \quad \text{(c)} \begin{pmatrix} a & 0 & 0 \\ 0 & b & 0 \\ 0 & 0 & c \end{pmatrix} \quad \text{with } a, b, c > 0.$$

$$\text{(d)} \begin{pmatrix} 6 & 2 & 1 \\ -4 & -1 & 4 \\ 4 & 11 & 0 \end{pmatrix} \quad \text{(e)} \begin{pmatrix} 0 & 2 & 1 \\ 1 & -3 & 3 \\ -1 & 4 & -5 \end{pmatrix}$$

Exercise 2.22
1. Each of players A and B chooses one of the numbers 1, 2, or 3. If the numbers are the same, A wins that amount from B. If the numbers differ, A loses to B the amount that A bet. Construct the game matrix and find an equilibrium pair and the value v.
2. The same problem as (1) except that, when the numbers differ, A loses to B the difference between the chosen numbers.

2.5   2 × 2 matrix games

Assume that player A has two strategies A_1 and A_2 and player B has two strategies B_1 and B_2. Let the game matrix have the form

$$\begin{pmatrix} a & b \\ c & d \end{pmatrix}$$

When the matrix has a saddle point, we know (by Exercise 2.17) that the pure strategies X and Y corresponding to that point form an equilibrium pair and the value of the game is the saddle point entry. So, from now on, if nothing is said to the contrary, we shall assume that the matrix has no saddle point.

If X is a mixed strategy for player A, then, since the coordinates of X are nonnegative real numbers that add up to 1, we can write $X = (x, 1 - x)$ where x is any real number such that $0 \leq x \leq 1$. Here, x is the probability that A plays pure strategy A_1 and $1 - x$ is the probability that A plays pure strategy A_2.

Likewise, we can represent any mixed strategy Y for player B in the form $Y = (y, 1 - y)$ with $0 \leq y \leq 1$. If players A and B play strategies X and Y, respectively, then the probability that A plays pure strategy A_1 and B plays pure strategy B_1 is xy (with corresponding pay-off a), the probability that A plays pure strategy A_1 and B plays pure strategy B_2 is $x(1 - y)$ (with corresponding pay-off b), and so on for the other two possibilities. Hence the expected pay-off $P(X, Y)$ for A is:

(1) $xya + x(1 - y)b + (1 - x)yc + (1 - x)(1 - y)d = xy(a - b - c + d) + x(b - d) + y(c - d) + d$

In order to find an equilibrium pair, it will turn out to be useful to represent (1) in the following form:

(2) $R(x - S)(y - T) + U$

where R, S, T, and U are suitable constants. Expanding (2), we get

(3) $Rxy - TRx - Sry + RST + U$

and, comparing the coefficients in (1) and (3), we obtain

(4) $R = a - b - c + d$ $TR = d - b$ $SR = d - c$ $RST + U = d$

Now note that $R \neq 0$. (If $R = 0$, then $a - b = c - d$ and, therefore, either $a \geq b$ and $c \geq d$ or $b \geq a$ and $d \geq c$. So, one column of the matrix would dominate the other; the matrix could be reduced to a single column, and then, again by dominance, to a single entry. That entry would be a saddle point of the matrix, contrary to hypothesis.) Solving for S, T, and U yields: $T = (d - b)/R$, $S = (d - c)/R$, and

$$U = d - \frac{(d - c)(d - b)}{R} = \frac{d(a + d) - d(b + c) - (d - c)(d - b)}{R}$$

$$= \frac{ad - bc}{R} = \begin{vmatrix} a & b \\ c & d \end{vmatrix} / R$$

These values of R, S, T, and U yield the desired representation:

(5) $P(X, Y) = R(x - S)(y - T) + U$.

It follows that, for $x = S$ and $y = T$, that is, when $X^* = (S, 1 - S)$ and $Y^* = (T, 1 - T)$, (X^*, Y^*) is an equilibrium pair. To see this, note that $P(X^*, Y) = U$ for any Y and $P(X, Y^*) = U$ for any X. Hence, $u(X^*) = U$ and $w(Y^*) = U$. So, (X^*, Y^*) is an equilibrium pair, the value v of the game is $P(X^*, Y^*) = U$, and the same pay-off v is obtained when X^* goes against any Y or when Y^* goes against any X. We can summarize these results in the following proposition.

Theorem 2.14

If the matrix

$$\begin{pmatrix} a & b \\ c & d \end{pmatrix}$$

of a game has no saddle point, then there is an equilibrium pair (X^*, Y^*), where $X^* = (S, 1 - S)$, $Y^* = (T, 1 - T)$ and $S = (d - c)/R$, $T = (d - b)/R$, $R = a - b - c + d$. (Note that R is the sum $a + d$ of the main diagonal entries minus the sum $b + c$ of the entries on the other diagonal.) The value v of the game is $(ad - bc)/R$. Moreover, if A plays X^*, the expected pay-off will be v, no matter what strategy B plays, and, if B plays Y^*, the expected pay-off also will be v, no matter what strategy A plays.

Example 2.18

Let us apply Theorem 2.14 to Matching Pennies (Example 2.13). The game matrix is

$$\begin{pmatrix} 1 & -1 \\ -1 & 1 \end{pmatrix}$$

$R = (1 + 1) - ((-1) + (-1)) = 4$, $S = (1 - (-1))/4 = \frac{1}{2}$, $T = (1 - (-1))/4 = \frac{1}{2}$, and $v = ((1)(1) - (-1)(-1))/4 = 0$. So, the equilibrium pair (X^*, Y^*) consists of $X^* = (\frac{1}{2}, \frac{1}{2})$ and $Y^* = (\frac{1}{2}, \frac{1}{2})$. Since $v = 0$, the game is fair.

Example 2.19

Let us use Theorem 2.14 to find equilibrium strategies for the game given by the 2×2 matrix

$$\begin{pmatrix} 8 & 6 \\ 4 & 7 \end{pmatrix}$$

There is no saddle point. $R = 15 - 10 = 5$, $S = (7 - 4)/5 = 3/5$, $T = (7 - 6)/5 = 1/5$, and $v = ((8)(7)-(6)(4))/5 = 32/5$. The equilibrium strategies X^* and Y^* are $(3/5,2/5)$ and $(1/5,4/5)$. Since $v = 32/5 > 0$, the game favors player A.

Exercise 2.23

Prove that, if the matrix

$$\begin{pmatrix} a & b \\ c & d \end{pmatrix}$$

has no saddle point, then $b \neq d$, $c \neq d$, $b \neq a$, and $c \neq a$. However, $a = d$ and $b = c$ are not excluded. (Hint: If $b = d$ or $c = a$, one row dominates the other and we get a saddle point. If $c = d$ or $b = a$, one column dominates the other and this yields a saddle point.)

Exercise 2.24

Show that the formulas for S and T in Theorem 2.14 give feasible values, and, in fact, $0 < S < 1$ and $0 < T < 1$. (Hint: Since there are no saddle points,

$d - c$ and $a - b$ have the same sign; for, if $d - c$ and $a - b$ had opposite signs, we would have a dominance relation and, thence, a saddle point. Similarly, $d - b$ and $a - c$ have the same sign. Now, $R = (a - b) + (d - c)$. So, if $d - c > 0$, then $a - b > 0$, $R > d - c > 0$ and $0 < S = (d - c)/R < 1$. If $d - c < 0$, then $a - b < 0$, $R < d - c$, and $0 < S = (d - c)/R < 1$. Use a similar argument to show that $0 < T < 1$.)

Exercise 2.25
Show that the equilibrium pair (X^*, Y^*) of Theorem 2.14 is unique. (Hint: Assume (X_1, Y_1) is another equilibrium pair, with $X_1 = (x_1, 1 - x_1)$ and $Y_1 = (y_1, 1 - y_1)$. Either $x_1 \neq S$ or $y_1 \neq T$. But $P(X_1, Y_1) = R(x_1 - S)(y_1 - T) + U$, $P(X_1, Y_1) = v = U$, and $R \neq 0$. Hence, we cannot have both $x_1 \neq S$ and $y_1 \neq T$. Consider the case where $x_1 = S$ and $y_1 \neq T$. Since, by Exercise 2.24, $0 < S < 1$, we can choose x_2 with $0 < x_2 < 1$ and $R(x_2 - S)(y_1 - T) > 0$. Let $X_2 = (x_2, 1 - x_2)$. Then $P(X_2, Y_1) = R(x_2 - S)(y_1 - T) + U > U$, contradicting the equilibrium pair requirement that $P(X_1, Y_1) = \max_{\text{all } X} P(X, Y_1)$. A similar argument leads to a contradiction when $x_1 \neq S$ and $y_1 = T$.)

Exercise 2.26
Find an equilibrium pair and the value for the games determined by the following matrices.

(a) $\begin{pmatrix} 90 & 95 \\ 91 & 89 \end{pmatrix}$

(Hint: Simplify the calculations by using Exercise 2.14(1).)

(b) $\begin{pmatrix} -1/3 & 4 \\ 1/2 & 0 \end{pmatrix}$

(Hint: Simplify the calculations by using Exercise 2.14(2) in order to work with integers only.)

(c) $\begin{pmatrix} 4 & 5 & -1 \\ 2 & 2 & 4 \\ 3 & 4 & 5 \end{pmatrix}$ (Hint: Look for dominance relations.)

(d) $\begin{pmatrix} 0 & -2 & -1 \\ 1 & 0 & 2 \\ -1 & 2 & 1 \end{pmatrix}$ (e) $\begin{pmatrix} 5 & 8 \\ 6 & 7 \end{pmatrix}$ (f) $\begin{pmatrix} -2 & 3 & 1 \\ -1 & 0 & 2 \\ 2 & 4 & 1 \end{pmatrix}$

Exercise 2.27

Melanie and Peter play the following card game. Melanie has two cards, a 4 and a 9. She puts one of these cards face down and Peter is supposed to guess it. If Peter guesses correctly, he wins from Melanie an amount equal to the number shown on the card. When Peter guesses incorrectly, he loses a certain fixed positive amount K to Melanie.

1. Set up a matrix for Melanie's winnings.
2. For what value(s) of K will there be a saddle point?
3. What should K be in order for the game to be fair?
4. When the game is fair, what are the optimal strategies for the players?

Exercise 2.28

Hilary and Craig independently and secretly choose either a Head or a Tail. If they both choose Heads, Hilary wins $4 from Craig. If they both choose Tails, Hilary wins $9 from Craig. If Hilary chooses Heads and Craig chooses Tails, then Craig wins $10 from Hilary. If Hilary chooses Tails and Craig chooses Heads, then Craig wins $5 from Hilary.

1. Find a matrix for the game.
2. What are optimal strategies for Hilary and Craig?
3. Is the game fair? What is its value?

Exercise 2.29
1. Let player A have strategies $A_1, ..., A_m$ and player B have strategies $B_1, ..., B_n$. If B plays the mixed strategy $Y = (y_1, ..., y_n)$, show that A can maximize his expected winnings by playing that pure strategy A_i for which $P(A_i, Y)$ is the maximum among $\{P(A_1, Y), ..., P(A_m, Y)\}$.
2. Given the game matrix

$$\begin{pmatrix} 3 & 1 & 0 \\ -1 & 2 & 1 \\ 1 & 3 & 2 \end{pmatrix}$$

if B plays the mixed strategy (1/6, 1/3, 1/2), what is the best (mixed or pure) strategy for A? For the same game matrix, if A plays the mixed strategy (1/2, 1/4, 1/4), what strategy should B use?

Exercise 2.30

There are three antibiotics A_1, A_2, A_3 for treating a certain bacterial disease, of which there are five different variations $B_1 - B_5$. The following game matrix has entries that are the probabilities that the antibiotics are effective against the particular kinds of bacteria.

$$\begin{pmatrix} .3 & .6 & .4 & .5 & 0 \\ 1 & 0 & 0 & 0 & 0 \\ 1 & .5 & 0 & 0 & 1 \end{pmatrix}$$

1. Find the optimal use of the antibiotics, assuming that Nature plays like a rational human being.
2. What treatment is best if one knows that the frequency of the bacterial diseases is (.1, .3, .3, .1, .2)?

Exercise 2.31

For the following matrix games with saddle points, find *all* the equilibrium pairs.

$$\text{(a)} \begin{pmatrix} 2 & 2 \\ 1 & 0 \end{pmatrix} \qquad \text{(b)} \begin{pmatrix} 2 & 2 \\ 1 & 5 \end{pmatrix} \qquad \text{(c)} \begin{pmatrix} 2 & 3 \\ 1 & 5 \end{pmatrix}$$

Exercise 2.32

Show that if

$$\begin{pmatrix} a & b \\ c & d \end{pmatrix}$$

has a saddle point, then its transpose

$$\begin{pmatrix} a & c \\ b & d \end{pmatrix}$$

also has a saddle point.

Exercise 2.33

Show that, if the 3 × 3 matrix

$$\begin{pmatrix} a & b & c \\ d & e & f \\ g & h & i \end{pmatrix}$$

has a saddle point, then its transpose

$$\begin{pmatrix} a & d & g \\ b & e & h \\ c & f & i \end{pmatrix}$$

need not have a saddle point.

Exercise 2.34
Consider the 2×2 matrix game with the matrix

$$\begin{pmatrix} 4 & -2 \\ 1 & 3 \end{pmatrix}$$

1. Find optimal mixed strategies for players A and B and find the value v of the game.
2. Assume that the game is played 80 times.
   (i) How much can A make sure that he will win?
   (ii) What is A's expected total gain if he plays his optimal strategy?
   (iii) What can B make sure will be his maximum total loss ?
   (iv) If B plays his optimal strategy, what total loss can he expect?

2.6 $2 \times n$, $m \times 2$, and 3×3 matrix games

Let us consider some simple examples to illustrate the methods used for $2 \times n$ and $m \times 2$ matrix games.

Example 2.20
Consider the 2×3 matrix

$$\begin{pmatrix} 2 & 4 & 3 \\ 4 & 0 & 1 \end{pmatrix}$$

There are no saddle points and no dominance relations. Let A_1 and A_2 be player A's strategies, and B_1, B_2, B_3 player B's strategies. Let $X = (x, 1 - x)$ be a mixed strategy for player A, with $0 \le x \le 1$. Then $u(X) = \min(P(X,B_1), P(X, B_2), P(X,B_3)) = \min(2x + 4(1 - x), 4x, 3x + (1 - x)) = \min (4 - 2x, 4x, 1 + 2x)$. The graphs of $y_1 = 4 - 2x$, $y_2 = 4x$, $y_3 = 1 + 2x$ on the interval $[0,1]$ are the line segments QR, OP, and ST in Figure 2.1.

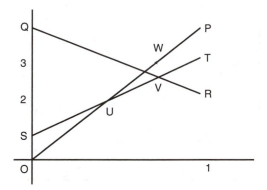

Figure 2.1

The graph OUVR of $u(X) = \min(y_1, y_2, y_3)$ consists of three line segments, OU, UV, and VR. Let (X^*, Y^*) be an equilibrium pair.[12] We know in general that, for an equilibrium pair, $u(X^*)$ is the maximin, that is, $u(X^*) = \max\limits_{0 \le x \le 1} u(X)$. From the graph we see that this maximum occurs at the point V, the intersection of QR and ST. By solving the equations $y = 4 - 2x$ and $y = 1 + 2x$ for QR and ST, we find that the x-coordinate at V is 3/4 and the y-coordinate at V, $u(X^*)$, the value v of the matrix game, is 5/2. Thus, $X^* = (^3/_4, {}^1/_4)$. Now we must find Y^*. Let $Y^* = (y_1, y_2, y_3)$. Note that V is the intersection of the lines $y = 4 - 2x$ and $y = 1 + 2x$, determined by $P(X, B_1)$ and $P(X, B_3)$. It is plausible then that strategy B_2 is not involved in the equilibrium strategy Y^*, that is, that $y_2 = 0$. Let us show that this is in fact so. At the x-coordinate $x = {}^3/_4$ of V, the point W on the line OP determined by $P(X, B_2)$ lies above V and, therefore, the y-coordinate $P(X^*, B_2)$ of W is greater than v, the y-coordinate at V. Hence, $y_2 = 0$, for, if $y_2 > 0$, then, by Theorem 2.12, $P(X^*, B_2)$ would be equal to v. Since $y_2 = 0$, the second column can be dropped from the game matrix, obtaining

$$\begin{pmatrix} 2 & 3 \\ 4 & 1 \end{pmatrix}$$

By the formula in Theorem 2.14 for the equilibrium pair of a 2×2 matrix without saddle points, we get $x = {}^3/_4$ (which we knew already) and $y_1 = {}^1/_2$. Therefore, $y_3 = {}^1/_2$. So, our equilibrium strategies are $X^* = (3/4, 1/4)$ and $Y^* = (1/2, 0, 1/2)$. Since $w(Y^*) = \max(P(A_1, Y^*), P(A_2, Y^*)) = \max(5/2, 5/2) = 5/2$ and $u(X^*) = 5/2$, it follows that $u(X^*) = w(Y^*)$ and that (X^*, Y^*) is actually an equilibrium pair.

---

[12]That there is such a pair follows from von Neumann's Theorem, which has not yet been proved. However, when we find (X^*, Y^*) and verify that $u(X^*) = w(Y^*)$, we will know that (X^*, Y^*) actually is an equilibrium pair.

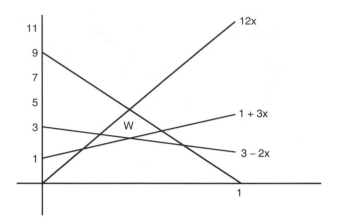

Figure 2.2

Example 2.21
Consider the game with the following 2×4 matrix.

$$\begin{pmatrix} 1 & 12 & 4 & 0 \\ 3 & 0 & 1 & 9 \end{pmatrix}$$

There are no dominance relations and no saddle points. Let A_1 and A_2 be player A's strategies and B_1, B_2, B_3, and B_4 player B's strategies. For any mixed strategy $X = (x, 1 - x)$ for A, $u(X) = \min(P(X, B_1), P(X, B_2), P(X, B_3), P(X, B_4)) = \min((1(x) + 3(1 - x), 12x + 0(1 - x), 4x + 1(1 - x), 0(x) + 9(1 - x)) = \min(3 - 2x, 12x, 1 + 3x, 9 - 9x)$. Figure 2.2 shows the line segments that are the graphs of $3 - 2x$, $12x$, $1 + 3x$, and $9 - 9x$ on the interval $[0, 1]$.

The graph of $u(X)$ is the minimum of the values on the four lines. If (X^*,Y^*) is an equilibrium pair, then $u(X^*)$ is attained at the maximum value of $u(X)$, that is, at the point W. Since W is the intersection of $y = 1 + 3x$ and $y = 3 - 2x$, solving those equations yields $x = 2/5$ as the x-coordinate of X^*, and $v = 11/5$ as the value of the game. Since those two lines correspond to the strategies B_3 and B_1, we can conclude, by an argument similar to that used in Example 2.20, that the second and fourth coordinates of Y^* are 0. So the game matrix reduces to

$$\begin{pmatrix} 1 & 4 \\ 3 & 1 \end{pmatrix}$$

By Theorem 2.14, the equilibrium pair of this matrix is $(2/5,3/5)$ and $(3/5,2/5)$. Hence, the equilibrium pair (X^*, Y^*) consists of $X^* = (2/5,3/5)$ and $Y^* = (3/5, 0, 2/5, 0)$. We can confirm this by verifying that $u(X^*) = w(Y^*)$.

Exercise 2.35
Find equilibrium strategies for the following matrix games.

(a) $\begin{pmatrix} 3 & 1 & 0 \\ 0 & 2 & 6 \end{pmatrix}$ (b) $\begin{pmatrix} 4 & 1 & -1 & 1 \\ 0 & 2 & 3 & 4 \end{pmatrix}$ (c) $\begin{pmatrix} -6 & -1 & 7 & 4 \\ 7 & -2 & -2 & -5 \end{pmatrix}$

In the previous two examples, the equilibrium pairs were unique. This is not always the case, as the following example shows.

Example 2.22
Consider the game matrix

$$\begin{pmatrix} 4 & 1 & 0 \\ 0 & 1 & 4 \end{pmatrix}$$

Let A_1 and A_2 be player A's strategies, and B_1, B_2, B_3 player B's strategies. Let $X = (x, 1 - x)$ be a mixed strategy for player A, with $0 \le x \le 1$. Then $u(X) = \min(P(X, B_1), P(X, B_2), P(X, B_3)) = \min(4x, 1, 4 - 4x)$. The graph of $u(X)$ is shown in Figure 2.3. It starts at the origin, moves up the line $y = 4x$ to W, continues on $y = 1$ to point Z, and then moves down the line $y = 4 - 4x$ to the x-axis at $x = 1$.

For any equilibrium pair (X^*, Y^*), the maximum value v of $u(X)$ is attained at $u(X^*)$. That maximum is clearly 1, on the segment WZ. W and Z are the intersections of $y = 1$ with $y = 4x$ and $y = 4 - 4x$, respectively, so that the x-coordinates of W and Z are $^1/_4$ and $^3/_4$. Hence, $X^* = (x, 1 - x)$ for any x such that $^1/_4 \le x \le ^3/_4$. On the interval $(1/4, 3/4)$, the y-values on the graphs corresponding to B_1 and B_3 are greater than v. Hence, by the reasoning in Example 2.20, the strategies B_1 and B_3 are not involved in Y^*, that is, $y_2 = 1$. Thus, there are infinitely many equilibrium pairs (X^*, Y^*), with $X^* = (x, 1 - x)$, $^1/_4 \le x \le ^3/_4$,

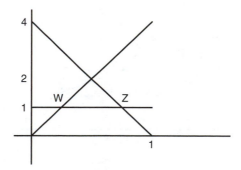

Figure 2.3

and $Y^* = (0, 1, 0)$. This is verified by noting that, for such (X^*, Y^*), $u(X^*) = 1 = w(Y^*)$.

Exercise 2.36
Find the equilibrium pairs for the following matrix games.

$$\text{(a)} \begin{pmatrix} 3 & 2 & 0 \\ 0 & 1 & 6 \end{pmatrix} \quad \text{(b)} \begin{pmatrix} 2 & 4 & 0 \\ 2 & 1 & 4 \end{pmatrix} \quad \text{(c)} \begin{pmatrix} 4 & 2 & 0 & 3 \\ 0 & 2 & 6 & 1 \end{pmatrix}$$

Example 2.23
Consider the game matrix

$$\begin{pmatrix} 2 & 0 & 1 \\ 1 & 5 & 3 \end{pmatrix}$$

Let A_1 and A_2 be player A's strategies, and B_1, B_2, B_3 player B's strategies. Let $X = (x, 1 - x)$ be a mixed strategy for player A, with $0 \le x \le 1$. Then $u(X) = \min(P(X, B_1), P(X, B_2), P(X, B_3)) = \min(1 + x, 5 - 5x, 3 - 2x)$. The graph of $u(X)$, shown in Figure 2.4, starts from $(0, 1)$, moves up the line $y = 1 + x$ to the point V and then down the line $y = 3 - 2x$ to the point $(1, 0)$.

The three lines $y = 1 + x$, $y = 3 - 2x$, and $y = 5 - 5x$ have a common intersection at $V = (2/3, 5/3)$. The maximum value of $u(X)$ clearly occurs at V. Hence, if (X^*, Y^*) is an equilibrium pair, then $X^* = (2/3, 1/3)$ and the value v of the game is $5/3$. Now, however, in order to find Y^*, we cannot eliminate any column of the matrix, since the three lines corresponding to B_1, B_2, and B_3 all intersect at V. Let $Y^* = (y_1, y_2, 1 - y_1 - y_2)$ and we have to allow for the possibility that all three coordinates are nonzero. By Theorem

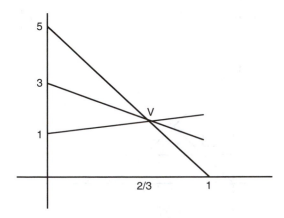

Figure 2.4

2.12, $P(A_1, Y^*) = P(A_2, Y^*) = v = 5/3$, that is, $2y_1 + (1 - y_1 - y_2) = 5/3$ and $y_1 + 5y_2 + 3(1 - y_1 - y_2) = 5/3$. Both of these equations reduce to $y_1 - y_2 = 2/3$. Since we also have $y_1 + y_2 \leq 1$, a simple calculation yields $2/3 \leq y_1 \leq 5/6$, $y_2 = y_1 - 2/3$, and $y_3 = (5/3) - 2y_1$. Hence, Y^* can be any of the infinitely many triples satisfying those three conditions. It is easy to see that $u(X^*) = 5/3 = w(Y^*)$ and, therefore, that these pairs (X^*, Y^*) are equilibrium pairs. (Note that Y^* ranges over all points on the line segment $(0, -2/3, 5/3) + y_1(1, 1, -2)$, where $2/3 \leq y_1 \leq 5/6$.)

Now let us look at some $m \times 2$ matrix games.

Example 2.24

1. Consider the game with the matrix:

$$\begin{pmatrix} 3 & -1 \\ 2 & 0 \\ -1 & 1 \end{pmatrix}$$

Let A_1, A_2, and A_3 be player A's strategies and B_1 and B_2 player B's strategies. Let $Y = (y, 1 - y)$ be a mixed strategy for B. Then $w(Y) = \max(3y - (1 - y), 2y, -y + (1 - y)) = \max(4y - 1, 2y, 1 - 2y)$. The graphs of $4y - 1$, $2y$, and $1 - 2y$ are shown in Figure 2.5 (with the horizontal axis as the y-axis).

The graph of $w(Y)$ starts at $(0, 1)$, moves down the graph of $1 - 2y$ to the point V, and then up segments of the graphs of $2y$ and $4y - 1$. If (X^*, Y^*) is an equilibrium pair, we know by Theorem 2.8(2) that $w(Y^*)$ is the minimum value of $w(Y)$, which is achieved at the point V. Therefore, if $Y^* = (y, 1 - y)$, then y is obtained by finding the intersection V of the graphs of $2y$ and $1 - 2y$. Setting $2y = 1 - 2y$, we find that $y = {}^1/_4$. Since the graph

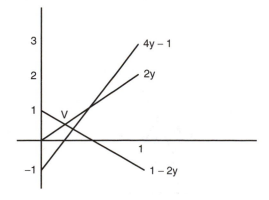

Figure 2.5

of $4y - 1$ lies below V when $y = \frac{1}{4}$, the strategy A_1 that is associated with $4y - 1$ will not be played by player A. Thus, the matrix can be reduced to

$$\begin{pmatrix} 2 & 0 \\ -1 & 1 \end{pmatrix}$$

By our results in Theorem 2.14 for 2×2 matrix games, the row strategies will be played with probabilities $\frac{1}{2}$ and $\frac{1}{2}$, and the column strategies with probabilities $\frac{1}{4}$ and $\frac{3}{4}$, and the value v is $\frac{1}{2}$. So, the resulting equilibrium pair for our game is $X^* = (0, \frac{1}{2}, \frac{1}{2})$ and $Y^* = (1/4, 3/4)$. That this is actually an equilibrium pair can be confirmed by verifying that $u(X^*) = \frac{1}{2} = w(Y^*)$.

2. Now let us look at the following 3×2 matrix game.

	B_1	B_2
A_1	1	1
A_2	4	0
A_3	0	4

Let $X^* = (x_1, x_2, x_3)$ and $Y^* = (y, 1 - y)$ form an equilibrium pair. Then $w(Y^*) = \max (1, 4y, 4 - 4y)$, with the V-shaped graph shown below, with vertex at V. From the graph we see that $v = \min w(Y^*) = 2$ at $y = .5$. So, $Y^* = (.5, .5)$. Since $P(A_1, Y^*) = 1 < v$, it follows that $x_1 = 0$. Hence the matrix reduces to

	B_1	B_2
A_2	4	0
A_3	0	4

So, by the formulas for 2×2 matrices, $x_2 = x_3 = .5$. Thus, $X^* = (0, .5, .5)$.

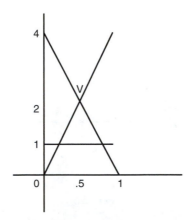

This correct solution of Example 2.24(2) exhibits in stark fashion the necessity for interpreting our results in the proper way. Let us assume that the units in the matrix of this example are millions of dollars. Since the value $v = 2$, player A will receive, in the long run, an average gain of 2 million dollars per game if he uses the mixed strategy of playing each of the strategies A_2 and A_3 with probability $1/2$ (and never playing strategy A_1). But that is only in the long run. What happens if player A is confronted with the problem of deciding what to do for just one play of this game? Then most people, put in this position, will choose strategy A_1, since that strategy guarantees a gain of one million dollars. If A obeys the optimal strategy $(0,1/2,1/2)$ and plays one of the strategies A_2 and A_3 at random (say, by flipping a fair coin), then A runs the risk (with probability $1/2$) of gaining nothing at all.

Exercise 2.37
Find all equilibrium pairs for the games given by the following matrices.

$$
\text{(a)} \begin{pmatrix} 2 & -1 \\ 1 & 1 \\ -1 & 2 \end{pmatrix}
\quad
\text{(b)} \begin{pmatrix} 1 & 4 \\ 2 & 0 \\ 0 & 5 \end{pmatrix}
\quad
\text{(c)} \begin{pmatrix} 3 & 0 \\ 2 & 4 \\ 0 & 6 \end{pmatrix}
$$

The following example is an often-cited instance of a concrete application of game-theoretic reasoning.

Example 2.25
(Davenport, [1960]) In a certain Jamaican fishing village, there are two fishing areas, the outer banks and the inner banks. The outer banks yield more and better fish, but swift currents may destroy the fish traps or the canoes. The inner banks are completely safe. The villagers use three strategies: (A_1) Set all traps on the inner banks; (A_2) Set all traps on the outer banks; (A_3) Set two-thirds of the traps on the outer banks. As pay-off, we use the average monthly profit in hundreds of dollars. We shall imagine that Nature is the opponent of the villagers and that Nature has two strategies: (B_1) Currents, and (B_2) No currents. Long experience justified the following table of expected pay-offs.

	B_1	B_2
A_1	17.3	11.5
A_2	− 4.4	20.6
A_3	5.2	17.0

There are no saddle points and no dominance relations. So, we have a game with the 3×2 matrix

$$
\begin{pmatrix} 17.3 & 11.5 \\ -4.4 & 20.6 \\ 5.2 & 17.0 \end{pmatrix}
$$

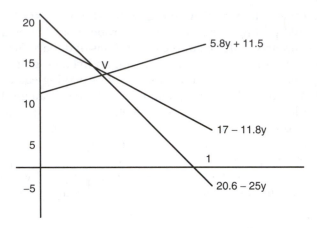

Figure 2.6

Let $Y = (y, 1 - y)$ be a mixed strategy for Nature. Then $w(Y) = \max(5.8y + 11.5, 20.6 - 25y, 17.0 - 11.8y)$. In Figure 2.6, the graphs of the components are shown.

The graph of $w(Y)$ consists of three line segments, with minimum value at point V, which is the intersection of the graphs of $5.8y + 11.5$ and $17 - 11.8y$. Setting $5.8y + 11.5 = 17 - 11.8y$, we get $y = 5/16$. So, at an equilibrium pair (X^*, Y^*), $Y^* = (5/16, 11/16)$. At $y = 5/16$, the graph of $20.6 - 25y$ lies below V and, therefore, strategy A_2 will be played with probability 0. Thus, our matrix reduces to the 2×2 matrix

$$\begin{pmatrix} 17.3 & 11.5 \\ 5.2 & 17.0 \end{pmatrix}$$

for which the equilibrium pair consists of (59/88,29/88) and (5/16,11/16), and the value v is $13 + (5/16)$. Hence, the equilibrium pair for the game is $X^* = (59/88,0,29/88)$ and $Y^* = (5/16,11/16)$. Note that 59/88 is roughly 2/3. Thus, the optimal strategy for the villagers is to set traps on the inner banks about two-thirds of the time and to set 2/3 of the traps on the outer banks about one-third of the time, and never to set all traps on the outer banks. In reality, out of 26 fishing canoes, on the average 18 used strategy A_1 and 8 used strategy A_3. So, the actual mixed strategy is (18/26, 0, 8/26), approximately (.69, 0, .31), which is close to (2/3, 0, 1/3). Our analysis had Nature running currents 5/16 of the time, whereas it turns out that currents run about $\frac{1}{4}$ of the time. Against Nature's actual strategy, strategy A_2 would yield the highest pay-off, about 14.35, as opposed to pay-offs of 14.05 and 12.95 for strategies A_3 and A_1. This casts doubt on the advisability of considering Nature as a player in a game.

Exercise 2.38
An automobile company buys a horn for one of its models. Normally it can buy the horn for $2, but, if the horn turns out to be defective, the actual cost to the

company will be \$8. However, at a fixed price of \$5, the supplier is willing to provide the horn and replace it at no extra cost if it is defective. Finally, the supplier offers to sign an "insurance" contract under which each horn costs \$7 but, if it is defective, the supplier replaces it and refunds the initial payment of \$7. Of these three plans, which should the company adopt? (Imagine that the company is playing against Nature, which sometimes causes a horn to be defective. Thus, Nature will have two strategies, Defective Horn or Good Horn.)

Now, let us look at some 3×3 matrix games and the methods used to solve them. Examples 2.16 and 2.17 already have illustrated two such methods (but note that the technique in Example 2.16 only applies to symmetric matrix games).

Example 2.26

Consider the game given by the 3×3 matrix

$$\begin{pmatrix} 4 & 1 & 3 \\ 5 & 2 & 1 \\ 3 & 4 & 4 \end{pmatrix}$$

Let $X = (x_1, x_2, x_3)$ and $Y = (y_1, y_2, y_3)$ form an equilibrium pair. Let us assume that each $y_j > 0$. Then, by Theorem 2.12, $P(X, B_j) = v$ for $j = 1, 2, 3$. Thus, $4x_1 + 5x_2 + 3x_3 = v$, $x_1 + 2x_2 + 4x_3 = v$, $3x_1 + x_2 + 4x_3 = v$

Eliminating v, we get $3x_1 + 3x_2 - x_3 = 0$ and $2x_1 - x_2 = 0$. Hence, $x_2 = 2x_1$ and $x_3 = 9x_1$. From $x_1 + x_2 + x_3 = 1$, we obtain $12x_1 = 1$. So, $x_1 = 1/12$, $x_2 = 2/12$, $x_3 = 9/12$. Similarly, by Theorem 2.12, $P(A_i, Y) = v$ for $i = 1, 2, 3$. Thus, $4y_1 + y_2 + 3y_3 = v$, $5y_1 + 2y_2 + y_3 = v$, $3y_1 + 4y_2 + 4y_3 = v$

Using $y_1 + y_2 + y_3 = 1$, we then obtain $y_1 = 7/12$, $y_2 = 1/12$, $y_3 = 4/12$. So, $X = (1/12, 2/12, 9/12)$ and $Y = (7/12, 1/12, 4/12)$. Note that $u(X) = \min(P(X, B_1), P(X, B_2), P(X, B_3)) = \min(41/12, 41/12, 41/12) = 41/12$ and $w(Y) = \max(P(A_1, Y), P(A_2, Y), P(A_3, Y)) = \max(41/12, 41/12, 41/12) = 41/12$. Thus, $u(X) = w(Y) = 41/12 = v$ and (X, Y) is an equilibrium pair.

Example 2.27

Consider the game given by the 3×3 matrix

$$\begin{pmatrix} 2 & -3 & 0 \\ 1 & 1 & 4 \\ 1 & 2 & 3 \end{pmatrix}$$

Let $X = (x_1, x_2, x_3)$ and $Y = (y_1, y_2, y_3)$ form an equilibrium pair. Let us assume that each $y_j > 0$. Then, by Theorem 2.12, $P(X, B_j) = v$ for $j = 1, 2, 3$. Thus, $2x_1 + x_2 + x_3 = v$, $-3x_1 + x_2 + 2x_3 = v$, $4x_2 + 3x_3 = v$.

Subtracting the third equation from the second, we get $-3x_1 - 3x_2 - x_3 = 0$. Since each $x_i \geq 0$, it follows that $x_1 = x_2 = x_3 = 0$, contradicting $x_1 + x_2 + x_3 = 1$. Hence, at least one $y_j = 0$.

Case 1. $y_1 = 0$. Then the game matrix reduces to the 3×2 matrix

$$\begin{pmatrix} -3 & 0 \\ 1 & 4 \\ 2 & 3 \end{pmatrix}$$

The entry 2 is a saddle point. So, we get $X = (0,0,1)$, $Y = (0,1,0)$. But, $u(X) = 1$ and $w(Y) = 2$. Hence, (X,Y) is not an equilibrium pair. So, $y_1 \neq 0$.

Case 2. $y_2 = 0$. The game matrix reduces to the 3×2 matrix

$$\begin{pmatrix} 2 & 0 \\ 1 & 4 \\ 1 & 3 \end{pmatrix}$$

The third row is dominated, yielding the 2×2 matrix

$$\begin{pmatrix} 2 & 0 \\ 1 & 4 \end{pmatrix}$$

By Theorem 2.14, $X = (1/5, 4/5, 0)$, $Y = (4/5, 0, 1/5)$. But, $u(X) = 1/5$ and $w(Y) = 8/5$. Thus, (X, Y) is not an equilibrium pair and $y_2 \neq 0$.

Case 3. $y_3 = 0$. The game matrix reduces to

$$\begin{pmatrix} 2 & -3 \\ 1 & 1 \\ 1 & 2 \end{pmatrix}$$

Since the second row is dominated, we get

$$\begin{pmatrix} 2 & -3 \\ 1 & 2 \end{pmatrix}$$

By Theorem 2.14, we obtain $X = (1/6,0,5/6)$, $Y = (5/6,1/6,0)$. Since $u(X) = 7/6 = w(Y)$, (X, Y) is an equilibrium pair and $v = 7/6$.

Example 2.28

Consider the game given by the 3×3 matrix

$$\begin{pmatrix} 5 & -3 & 2 \\ 1 & 4 & 0 \\ 3 & 2 & 6 \end{pmatrix}$$

Let $X = (x_1, x_2, x_3)$ and $Y = (y_1, y_2, y_3)$ form an equilibrium pair. Let us assume that each $y_j > 0$. Then, by Theorem 2.12, $P(X, B_j) = v$ for $j = 1, 2, 3$. Thus, $5x_1 + x_2 + 3x_3 = v$, $-3x_1 + 4x_2 + 2x_3 = v$, $2x_1 + 6x_3 = v$

Solving these equations together with $x_1 + x_2 + x_3 = 1$, we get $x_1 = 8/52$, $x_2 = 27/52$, $x_3 = 17/52$. Since each $x_i > 0$, Theorem 2.12 yields $P(A_i, Y) = v$ for $i = 1, 2, 3$. So, $5y_1 - 3y_2 + 2y_3 = v$, $y_1 + 4y_2 = v$, $3y_1 + 2y_2 + 6y_3 = v$ from which we get $4y_1 - 7y_2 + 2y_3 = 0$ and $-2y_1 + 2y_2 - 6y_3 = 0$. Adding twice the second equation to the first yields $-3y_2 - 10y_3 = 0$. Since all $y_i \geq 0$, this implies that $y_2 = y_3 = 0$. Then the earlier equation $4y_1 - 7y_2 + 2y_3 = 0$ implies $y_1 = 0$, contradicting $y_1 + y_2 + y_3 = 1$. Hence, at least one $y_i = 0$. By reasoning like that in the previous example, we can eliminate the cases $y_1 = 0$ and $y_2 = 0$. Hence, $y_3 = 0$ and the game matrix reduces to

$$\begin{pmatrix} 5 & -3 \\ 1 & 4 \\ 3 & 2 \end{pmatrix}$$

There are no saddle points or dominance relations. So, we must employ the graphical technique used in Example 2.24. We leave it to the reader to show that we get $X = (0, 1/4, 3/4)$ and $Y = (1/2, 1/2, 0)$, and that $v = 5/2$.

Exercise 2.39
Find the equilibrium pairs and values for the following matrix games.

(a) $\begin{pmatrix} 3 & 4 & 2 \\ 1 & 5 & 3 \\ 2 & 1 & 2 \end{pmatrix}$ (b) $\begin{pmatrix} 1 & 3 & 4 \\ 2 & 4 & 2 \\ 0 & 1 & 3 \end{pmatrix}$ (c) $\begin{pmatrix} -1 & 2 & 0 \\ 3 & 1 & 1 \\ 1 & 0 & 2 \end{pmatrix}$ (d) $\begin{pmatrix} 1 & 2 & 3 \\ 3 & 2 & 1 \\ 1 & 3 & 2 \end{pmatrix}$

(e) $\begin{pmatrix} 3 & 1 & 5 \\ 2 & 4 & 1 \\ 4 & 3 & 2 \end{pmatrix}$

2.7 Linear programming

A *linear programming problem* (lpp) requires us to find the maximum (or minimum) value of a linear function $f(x_1, \ldots, x_n) = c_1 x_1 + c_2 x_2 + \cdots + c_n x_n$ if the variables $x_1, \ldots, x_n$ are restricted to be nonnegative and to satisfy some linear inequalities:

$$a_{11}x_1 + a_{12}x_2 + \cdots + a_{1n}x_n \leq b_1$$

$$a_{21}x_1 + a_{22}x_2 + \cdots + a_{2n}x_n \leq b_2$$

$$\cdots\cdots\cdots\cdots\cdots\cdots\cdots$$

$$a_{m1}x_1 + a_{m2}x_2 + \cdots + a_{mn}x_n \leq b_m$$

Note
1. Any of the $\leq$ linear inequalities above could be replaced by an equiv-alent $\geq$ linear inequality. For example, $2x_1 - 5x_2 \leq 3$ is equivalent to $-2x_1 + 5x_2 \geq -3$. Thus, a linear programming problem may involve only $\leq$ inequalities, only $\geq$ inequalities, or any combination of $\leq$ and $\geq$ inequalities.
2. Linear equations also are permitted, since such equations are equiv-alent to pairs of inequalities. For example, $4x_1 + x_2 = 3$ is equivalent to the conjunction of the two inequalities $4x_1 + x_2 \leq 3$ and $4x_1 + x_2 \geq 3$.

Example 2.29
Consider the following lpp's. Find the maximum and minimum values of $x_1 + 2x_2$ subject to the conditions

$$x_1 \geq 0, x_2 \geq 0$$

$$2x_1 + 3x_2 \leq 12$$

$$-x_1 + x_2 \leq 1$$

These inequalities require that (x_1, x_2) lies on or inside the quadrilateral OABC in Figure 2.7, that is, (x_1, x_2) is in the first quadrant, under the line $2x_1 + 3x_2 = 12$ and under the line $-x_1 + x_2 = 1$.

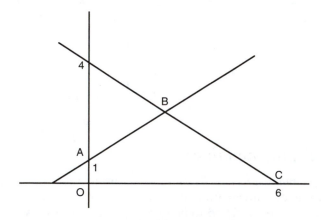

Figure 2.7

The function to be maximized $f(x_1, x_2) = x_1 + 2x_2$ is continuous and the domain D, the quadrilateral OABC and its interior, is closed and bounded. Therefore, there must be maximum and minimum values of f on D. But we are even more fortunate, because, in all such cases of lpp's, the maximum and minimum values must be attained at one of the vertices (in this example, O, A, B, or C). This result, called the Vertex Theorem, will be discussed below. In this example, the values of f at O, A, B, C are 0, 2, 37/5, 6 (since O = (0, 0), A = (0, 1), B = (9/5, 14/5), C = (6, 0)). Thus, f attains its maximum 37/5 when $x_1 = 9/5$ and $x_2 = 14/5$ and its minimum 0 when $x_1 = x_2 = 0$.

In general, for any lpp, the function $f(x_1, ..., x_n)$ to be maximized or minimized is called the *objective function*, the inequalities or equalities that must be satisfied are called the *constraints*, any point that satisfies the constraints is called a *feasible point*, and the set of feasible points is called the *constraint set*.

If the objective function is to be maximized and it actually has a maximum, then that maximum is called the *value* of the lpp. Similarly, if the objective function is to be minimized and it actually has a minimum, then that minimum is called the *value* of the lpp.

Note that any point $(u_1, ..., u_n)$ in the constraint set at which $f(x_1, ..., x_n)$ achieves a maximum is a point at which the function $g(x_1, ..., x_n) = -f(x_1, ..., x_n)$ achieves a minimum. Hence, a maximization problem can be replaced by an equivalent minimization problem, and vice versa.

Unless something is said to the contrary, we shall assume in any lpp that the variables are taken to be nonnegative. From now on, when we refer to the *constraints* of an lpp, we do not mean the implicit constraints $x_1 \geq 0$, $x_2 \geq 0, ..., x_n \geq 0$.

The Vertex Theorem, which was applied above in Example 2.29, asserts that a maximum or minimum of an objective function $f(x_1, x_2)$ on a bounded constraint set will occur at a vertex of the constraint set. To see this, consider any line segment that is the intersection of a line with the constraint set. Let $f(x_1, x_2) = r_1 x_1 + r_2 x_2$ and let the line segment have endpoints (a, b) and (c, d). The segment can be given by the parametric equations $x_1 = (c - a)t + a$ and $x_2 = (d - b)t + b$ for $0 \leq t \leq 1$. So, $f(x_1, x_2) = r_1 x_1 + r_2 x_2 = r_1((c - a)t + a) + r_2((d - b)t + b) = Mt + N$, where $M = r_1(c - a) + r_2(d - b)$ and $N = r_1 a + r_2 b$.

If $M > 0$, $f(x_1, x_2)$ is an increasing function of t and its minimum and maximum on the segment occur at (a, b) and (c, d), respectively. If $M < 0$, $f(x_1, x_2)$ is a decreasing function of t and its minimum and maximum on the segment occur at (c, d) and (a, b), respectively. If $M = 0$, $f(x_1, x_2)$ is constant on the segment and its identical minimum and maximum occur at (a, b) and (c, d), as well as all interior points. These results yield the Vertex Theorem. (The reader should fill in the further details of the argument.) The theorem can be generalized to assert that, if the objective function has a minimum or a maximum, then that optimal value will occur at a vertex. The Vertex Theorem can be extended to higher dimensions, but we will not pursue the matter any further here.

Exercise 2.40
Verify the assertion above that $f(x_1,...,x_n)$ achieves a maximum at the point $(u_1,...,u_n)$ if and only if $-f(x_1,...,x_n)$ achieves a minimum at $(u_1,...,u_n)$.

Exercise 2.41
Solve the following lpp's. (In each case, draw a diagram, find the constraint set, and apply the Vertex Theorem.)

1. Find the maximum and minimum values of $f(x_1, x_2) = 5x_1 - 3x_2$, subject to the constraints $2x_1 - 3x_2 \le 6$, $x_1 + 5x_2 \le 10$.
2. Find the maximum and minimum values of $f(x_1, x_2) = 6x_1 + x_2$, subject to the constraints $7x_1 + 2x_2 \le 14$, $4x_1 + 3x_2 \le 12$.
3. Find the maximum and minimum values of $f(x_1, x_2) = 3x_1 - 5x_2$, subject to the constraints $x_1 + x_2 \le 5$, $2x_1 + 3x_2 \ge 6$, $4x_1 - x_2 \le 4$.
4. Find the maximum and minimum values of $f(x_1, x_2) = 3x_1 - x_2$, subject to the constraints $2x_1 + 3x_2 \le 6$, $4x_1 + 3x_2 = 8$.

Example 2.30
Consider the following pair of lpp's: Find the maximum and minimum of $f(x_1,x_2) = x_2 - x_1$, subject to the constraints $x_1 - 2x_2 \le 0$ and $3x_1 - 2x_2 \ge 0$. The constraint set consists of all points in the first quadrant on or between the lines through the origin with slopes $1/2$ and $3/2$. (In Figure 2.8, this is the set of all points on or within the angle $\angle AOB$, where OA is the line $3x_1 - 2x_2 = 0$ and OB is the line $x_1 - 2x_2 = 0$.)

On the line OB, $x_2 = (3/2)x_1$, so that, for points (x_1, x_2) on that line, $f(x_1, x_2) = x_2 - x_1 = (1/2)x_1$. As a point moves away from O along the vector OB, x_1 approaches $+\infty$ and, therefore, $f(x_1, x_2)$ also approaches $+\infty$. Thus, f has no maximum value on the constraint set. On the line OA, $x_2 = (1/2)x_1$, so that, for points on that line, $f(x_1, x_2) = x_2 - x_1 = -(1/2)x_1$. As a point moves away

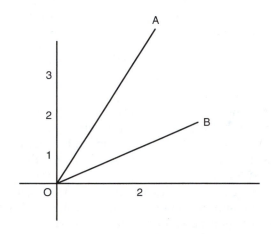

Figure 2.8

from O along the vector OA, x_1 approaches $+\infty$ and, therefore, $f(x_1,x_2)$ approaches $-\infty$. Thus, f has no minimum value on the constraint set. This example shows that lpp's need not have a solution.

Another way in which a lpp may fail to have a solution occurs when the constraint set is empty. For example, if the constraint set consists of all (x_1, x_2) in which x_1 and x_2 are nonnegative and satisfy the constraint $x_1 + x_2 \leq -1$, then the constraint set contains no points at all.

Exercise 2.42
Solve the following lpp's or show that no solution exists. (In each case, first find the constraint set. Remember that the variables are assumed to have nonnegative values.)

1. Maximize $f(x_1,x_2) = x_1 + x_2$ subject to the constraints $x_1 + x_2 \geq 1$ and $-x_1 + x_2 \leq 0$.
2. The same problem as (1) except that f has to be minimized.
3. Maximize $3x_1 + 7x_2$ subject to the constraints $x_1 + 2x_2 \geq 2$ and $2x_1 + 5x_2 \leq 10$.
4. The same problem as (3) except that the objective function is to be minimized.
5. The same problem as (3) except that the constraint $2x_1 + 5x_2 \leq 10$ is replaced by $2x_1 - 5x_2 \leq 10$.
6. The same problem as (3) except that $\leq$ and $\geq$ are interchanged in the constraints.
7. The same problem as (3) except that $2x_1 + 5x_2 \leq 10$ is replaced by $2x_1 + 5x_2 = 10$.

By a *standard maximization lpp* we mean an lpp in which the objective function is to be maximized and all the constraints are $\leq$ inequalities. By a *standard minimization lpp* we mean an lpp in which the objective function is to be minimized and all the constraints are $\geq$ inequalities.

Example 2.31
1. Maximize $u = 4x - 5y$ subject to the constraints

$$-x + 3y \leq 10$$
$$x + y \leq 8$$
$$-x - 2y \leq 9$$

This is a standard maximization lpp.
2. Minimize $v = 10a + 8b + 9c$ subject to the constraints

$$-a + b - c \geq 4$$
$$3a + b - 2c \geq -5$$

This is a standard minimization lpp.

The lpp's (1) and (2) in Example 2.31 are examples of what are called *dual* lpp's. First of all, there are two variables and three constraints in (1) and there are three variables and two constraints in (2).

Second, notice that the coefficients 4 and −5 of the variables x and y in the objective function u in (1) are the constants on the right-hand sides of the constraints in (2), and the coefficients 10, 8, and 9 of the variables a, b, and c in the objective function v in (2) are the constants on the right-hand sides of the constraints in (1).

Third, the coefficients −1 and 3 of the variables x and y in the first constraint of (1) are the coefficients of the first variable a in the two constraints of (2), the coefficients 1 and 1 of the variables x and y in the second constraint of (1) are the coefficients of the second variable b in the two constraints of (2), and the coefficients −1 and −2 of the variables x and y in the third constraint of (1) are the coefficients of the third variable c in the two constraints of (2). Similarly, the coefficients of each of the two constraints in (2) are the coefficients of the corresponding variable in the three constraints of (1).

In general, a standard maximization lpp and a standard minimization lpp are said to be *duals* of each other if:

First, the number of variables in each lpp is equal to the number of constraints in the other lpp; Second, the coefficients of the variables in the objective function of each lpp are equal to the constants on the right-hand side of the constraints of the other lpp; Third, the coefficients of the variables in each constraint (say, the *j*th constraint) of one lpp are equal to the coefficients of the corresponding variable (the *j*th variable) in the constraints of the other lpp.

Notice that, given a standard maximization or minimization lpp, one can write down a corresponding standard minimization or maximization lpp that is a dual of the given lpp.

Example 2.32

Consider the following standard maximization lpp.

Maximize $u = 3x - 5y$ subject to the constraints

$$x + 2y \leq 4,$$

$$3x + 5y \leq 7,$$

$$2x - 3y \leq 6$$

Since there are three constraints, the dual must have three variables, say a, b, and c. The dual's objective function, to be minimized, has the form $v = 4a + 7b + 6c$, since 4, 7, and 6 are the constants on the right-hand sides of the constraints of the given lpp. The dual must have two constraints, since the given lpp has two variables. Those constraints are

$$a + 3b + 2c \geq 3,$$

$$2a + 5b - 3c \geq -5$$

Note that these are $\geq$ inequalities; the constants 3 and -5 on the right-hand sides are the coefficients of the variables x and y in the objective function u of the given lpp; the coefficients 1, 3, and 2 of x in the constraints in the given lpp have become the coefficients of the variables a,b,c in the new first constraint, and the coefficients 2, 5, and -3 of y in the constraints in the given lpp have become the coefficients of the variables a, b, c in the new second constraint. Thus, the dual is: Minimize $v = 4a + 7b + 6c$ subject to the constraints

$$a + 3b + 2c \geq 3,$$
$$2a + 5b - 3c \geq -5$$

Example 2.33

Consider the following standard minimization lpp.

Minimize $v = 6a - 3b - c$ subject to the constraints

$$5a + 4b + 2c \geq 1,$$
$$-a + b + 4c \geq -4,$$
$$2a - 5b + 6c \geq 0$$

Let us find its dual. Since the given lpp has three constraints, the dual must have three variables, say x, y, and z. The objective function to be maximized must be $u = x - 4y$, since 1, -4, and 0 are the constants on the right-hand sides of the constraints of the given lpp. The dual will have three $\leq$ constraints, since the given lpp has three variables, and those constraints are

$$5x - y + 2z \leq 6,$$
$$4x + y - 5z \leq -3,$$
$$2x + 4y + 6z \leq -1$$

The constants on the right-hand sides 6, -3, and -1 are the coefficients of a, b, and c in the given objective function. The coefficients 5, -1, 2 in the first of these constraints are the coefficients of the first variable a in the constraints of the given lpp; the coefficients 4, 1, -5 of the second of these constraints are the coefficients of the second variable b in the constraints of the given lpp; the coefficients 2, 4, 6 of the third of these constraints are the coefficients of the third variable c in the constraints of the given lpp.

Here is a more precise definition of dual lpp's. The following standard maximization and standard minimization lpp's (1) and (2) are defined to be duals of each other.

1. Maximize $\sum_{j=1}^{n} c_j x_j + d$, subject to the constraints $\sum_{j=1}^{n} a_{ij} x_j \leq b_i, 1 \leq i \leq m$
2. Minimize $\sum_{i=1}^{m} b_i y_i + d$, subject to the constraints $\sum_{i=1}^{m} a_{ij} y_i \geq c_j, 1 \leq j \leq n$.

Exercise 2.43

In each of the following, find the dual of the given standard maximization or minimization lpp.

1. Maximize $u = 3x + 9y - z$ subject to the constraints $x - y + z \leq 2$, $3x + 3y - 2z \leq -4$, $2x - 3y - z \leq 5$, $x + y + 2z \leq 4$.
2. Minimize $v = 3a - 4b$ subject to the constraints $2a + 5b \geq 7$, $-4a - b \geq 0$.
3. Maximize $u = 2x - y - z + 4w$ subject to the constraints $3x + y - 2z - w \leq 3$, $7x - 4y + w \leq -2$.

A basic result of the theory of linear programming is the following proposition.

The fundamental theorem of duality

Given a pair of dual lpp's, if one of them has a solution, then so does the other, and the values of the two lpp's are equal.

A proof via the so-called simplex method is sketched in Chapter 3.

We want to use linear programming to obtain the following essential result. (The reader should review the material of Section 4.)

The fundamental theorem of matrix games (Von Neumann's Theorem)

For every matrix game, there exists an equilibrium pair (Von Neumann [1928]).

We shall show how this follows from the Fundamental Theorem of Duality. Let $C = (c_{ij})$ be an $m \times n$ game matrix. We must prove the existence of an equilibrium pair. In fact, the proof in Chapter 3 gives instructions as to how to find such a pair. Some examples will be worked out in Chapter 3.

Key point 1

We may assume all entries c_{ij} are positive.

To see this, note that we can add a suitable constant K to all entries in our matrix to obtain a matrix with only positive entries. By Exercise 2.14(1), if the new matrix has an equilibrium pair, then that pair also is an equilibrium pair of the original matrix (and the value of the original game is K less than the value of the new game). Thus, it suffices to derive von Neumann's Theorem just for games whose matrices have only positive entries.

Key point 2

To find an equilibrium pair, it will suffice to find mixed strategies X and Y and a number v such that the following two conditions hold:

1. $v \leq P(X, B_j)$ for $1 \leq j \leq n$, where B_j is the jth strategy for player B;
2. $v \geq P(A_i, Y)$ for $1 \leq i \leq m$, where A_i is the ith strategy for player A.

To see this, note that, from (1) it follows that $v \leq \min_j P(X, B_j) = u(X)$, and from (2) it follows that $v \geq \max_i P(A_i, Y) = w(Y)$. So, $w(Y) \leq v \leq u(X)$. On the other hand, by Theorem 2.6, $u(X) \leq w(Y)$ holds for any X and Y. Hence, $u(X) = w(Y) = v$. Therefore, by Theorem 2.7, (X, Y) would be an equilibrium pair, and $v = P(X, Y)$ would be the value of the game. Since all the entries c_{ij} are assumed to be positive, the value v must also be positive.

Conditions (1) and (2) of Key Point 2 can be written as two systems of linear inequalities:

1. $v \leq \displaystyle\sum_{i=1}^{m} c_{ij} x_i \quad 1 \leq j \leq n$

2. $v \geq \displaystyle\sum_{j=1}^{n} c_{ij} y_j \quad 1 \leq i \leq m$

If we divide all these inequalities by v, we obtain:

$$1 \leq \sum_{i=1}^{m} c_{ij}(x_i/v) \quad 1 \leq j \leq n$$

$$1 \geq \sum_{j=1}^{n} c_{ij}(y_j/v) \quad 1 \leq i \leq m$$

Let us introduce new variables $\alpha_i = x_i/v$ and $\beta_j = y_j/v$. So, the two systems of inequalities become:

$$1 \leq \sum_{i=1}^{m} c_{ij}\alpha_i \quad 1 \leq j \leq n$$

$$1 \geq \sum_{j=1}^{n} c_{ij}\beta_j \quad 1 \leq i \leq m$$

Note that

$$\sum_{i=1}^{m} \alpha_i = \sum_{i=1}^{m} (x_i/v) = (1/v) \sum_{i=1}^{m} x_i = 1/v, \text{ since } \sum_{i=1}^{m} x_i = 1$$

Likewise,

$$\sum_{j=1}^{n} \beta_j = \sum_{j=1}^{n} (y_j/v) = (1/v) \sum_{j=1}^{n} y_j = 1/v, \text{ since } \sum_{j=1}^{n} y_j = 1$$

Now, in the system (1), the value of v can be taken to be the maximum of all v's satisfying the inequalities (1). Since $1/v = \sum_{i=1}^{m} \alpha_i$, maximizing v

would be equivalent to minimizing $\sum_{i=1}^{m} \alpha_i$. Thus, we have been lead to the following lpp:

$$\text{(I*)} \quad \text{Minimize} \sum_{i=1}^{m} \alpha_i$$

subject to the constraints

$$\sum_{i=1}^{m} c_{ij}\alpha_i \geq 1 \qquad 1 \leq j \leq n$$

Similarly, in system (2), the value v can be taken to be the minimum of all v's satisfying the inequalities (2). Since $1/v = \sum_{j=1}^{n} \beta_j$, minimizing v is equivalent to maximizing $\sum_{j=1}^{n} \beta_j$. Hence, we want to solve the following lpp:

$$\text{(II*)} \quad \text{Maximize} \sum_{j=1}^{n} \beta_j$$

subject to the constraints

$$\sum_{j=1}^{n} c_{ij}\beta_j \leq 1 \qquad 1 \leq i \leq m$$

Notice the very happy circumstance that (I*) and (II*) are dual lpp's. In addition, since all entries c_{ij} are positive, the constraint set of (II*) is a non-empty closed, bounded subset of R^n, and, since $\sum_{j=1}^{n} \beta_j$ is a continuous function, that function will attain a maximum on the constraint set. Thus, (II*) has a solution. Therefore, by the Fundamental Theorem of Duality, (I*) also has a solution and the values of (I*) and (II*) are equal. Let us show how this leads to an equilibrium pair.

Let $\alpha_1^*, ..., \alpha_m^*$ and $\beta_1^*, ..., \beta_n^*$ be solutions of (I*) and (II*), respectively. In addition, the minimum value $\sum_{i=1}^{m} \alpha_i^*$ of (I*) and the maximum value $\sum_{j=1}^{n} \beta_j^*$ of (II*) are equal. Since the constraints $\sum_{i=1}^{m} c_{ij}\alpha_i^* \geq 1$ are valid and all c_{ij} are positive, at least one α_i^* must be positive, and, of course, all α_i^* are nonnegative. Hence, $\sum_{i=1}^{m} \alpha_i^*$ is positive. Let v be the reciprocal of $\sum_{i=1}^{m} \alpha_i^*$. Define $x_i^* = v\alpha_i^*$ and $y_j^* = v\beta_j^*$. Therefore, $\sum_{i=1}^{m} x_i^* = \sum_{i=1}^{m} v\alpha_i^* = v \sum_{i=1}^{m} \alpha_i^* = 1$. Thus, $X = (x_1^*, ..., x_m^*)$ is a mixed strategy for player A. Similarly, $\sum_{j=1}^{n} y_j^* = \sum_{j=1}^{n} v\beta_j^* = v \sum_{j=1}^{n} \beta_j^* = v \sum_{i=1}^{m} \alpha_i^* = 1$ and, therefore, $Y = (y_1^*, ..., y_n^*)$ is a mixed strategy for player B. From the constraints of (I*), $\sum_{i=1}^{m} c_{ij}\alpha_i^* \geq 1$ for $1 \leq j \leq n$, multiplication by v yields:

1. $$\sum_{i=1}^{m} c_{ij}x_i^* \geq v \qquad \text{for } 1 \leq j \leq n$$

Likewise, from the constraints of (II*), $\sum_{j=1}^{n} c_{ij}\, \beta_j^* \leq 1$ for $1 \leq i \leq m$, and multiplication by v yields:

2. $\displaystyle\sum_{j=1}^{n} c_{ij}\, y_j^* \leq v$ for $1 \leq i \leq m$

Thus, by Key Point 2, (X^*, Y^*) is an equilibrium pair.

This argument has shown how von Neumann's Theorem follows from the Fundamental Theorem of Duality. The proof of the latter in Chapter 3 produces a concrete method for finding the solutions $\alpha_1^*,\ldots, \alpha_m^*$ and $\beta_1^*,\ldots,$ β_n^* and then the argument above yields the desired equilibrium pair. Some relatively simple examples will be completely solved in Chapter 3. However, the calculation of the solutions of real-life problems usually requires so many steps and involves such large numbers that suitable computer programs have to be used. Solutions also can be obtained on the Internet, for example by the Matrix Game Solver.

chapter three

The simplex method.
The fundamental theorem
of duality. Solution of
two-person zero-sum games[1]

3.1 Slack variables. Perfect canonical linear programming problems

Equations are easier to deal with than inequalities. In linear programming problems, we can replace an inequality $a_1x_1 + a_2x_2 + \cdots + a_nx_n \leq b$ by the equivalent equality $a_1x_1 + a_2x_2 + \cdots + a_nx_n + r = b$ where r is a new variable. The equivalence depends on the fact that, in linear programming problems, variables are assumed to take only non-negative values. The new variable r is called a *slack variable*.

Example 3.1
$3x - 4y + 2z \leq 7$ is equivalent to $3x - 4y + 2z + r = 7$.

If we start with a standard maximization lpp and introduce distinct slack variables in each constraint, we call the resulting equivalent lpp a *slack system*.

Example 3.2
Consider the standard maximization lpp:

$$\text{Maximize } 4x - 2y + z + 5$$

subject to the constraints

$$3x + y - z \leq 4$$

$$x - 2y + 4z \leq -2$$

[1]The essential ideas in our presentation follow the beautiful exposition in Sultan [1993].

Introducing slack variables r and s, we obtain the slack system:

$$\text{Maximize } 4x - 2y + z + 5$$

subject to the constraints

$$3x + y - z + r = 4$$
$$x - 2y + 4z + s = -2$$

Slack systems are examples of the more general class of *canonical lpp's*. Every such lpp is a maximization lpp such that each constraint is an equation with a *distinguished* variable that has coefficient 1 and occurs in no other constraint. These distinguished variables are called *basic*; the other variables are said to be *non-basic*.

Example 3.3
In the slack system of Example 3.2, r and s are basic, and x, y, and z are non-basic.

Example 3.4
Consider the following maximization lpp.

$$\text{Maximize } 4x - 2y + 3z$$

subject to the constraints

$$3x + y - z + r = 4$$
$$2x - 3z + s = 7$$

This can be considered a canonical lpp in two different ways, first with basic variables r and s, and second with basic variables y and s. In general, the specification of a canonical lpp must tell us which variables are basic. Note that the objective function always can be written with the non-basic variables as its independent variables. In this example, the objective function $4x - 2y + 3z$ is already written in terms of the non-basic variables x, y, and z in the first case. In the second case, when x, z, and r are the non-basic variables, we can solve the first constraint equation to get $y = 4 - 3x + z - r$, so that the objective function can be rewritten as $4x - 2(4 - 3x + z - r) + 3z = 10x + z + 2r - 8$.

Exercise 3.1
In a canonical lpp with m constraints and n variables, show that $m \leq n$.

Given a canonical lpp, its *basic point* is defined by setting all non-basic variables equal to zero and then solving the constraint equations for the basic variables. By a *feasible basic point* we mean a basic point in which the values of the basic variables are non-negative. Since the values of the non-basic variables of a basic point are zero, a basic point is feasible if and only if all its coordinates are non-negative. If the constant of each constraint equation occurs, as usual, alone on the right side of the equation, then the feasibility of the basic point means that all those constants are non-negative. A feasible basic point is feasible in the original sense, since the constraints will be automatically satisfied. It will turn out that, if an lpp has a solution, it is attained at a feasible basic point of a suitable equivalent canonical lpp.

In the slack system of Example 3.2, the basic point is $(0,0,0,4,-2)$. Note that we arbitrarily arranged the variables in the order x, y, z, r, s. In general, the order of the coordinates of the basic point depends on an initial arrangement of the variables. In Example 3.4, if r and s are the basic variables and the variables are arranged in the order x, y, z, r, s, then the basic point is $(0,0,0,4,7)$; if the variables are arranged in the same order but y and s are the basic variables, then the basic point is $(0,4,0,0,7)$.

By a *perfect canonical lpp* we mean a canonical lpp for which the basic point is feasible and the objective function depends only on non-basic variables.

The slack system of Example 3.2 is not perfect, since the basic point $(0,0,0,4,-2)$ is not feasible. Of the two canonical lpp's of Example 3.4, the one with r and s as basic variables is perfect; but the one with y and s as basic variables is not perfect, since the objective function $4x - 2y + 3z$ depends on the variable y, which is basic.

Exercise 3.2

For the following standard maximization lpp's with the same objective function $6x - y + 3z$ and the indicated constraints, which can be considered perfect canonical lpp's?

1. Constraints $3x + 2y + r = 5$ and $x - 3z + s = 0$
2. Constraints $x - 3y - z + 2r = 3$ and $2x + y + s = 3$
3. Constraints $2x - y - z + r = -7$ and $x + 2y + z + s = 2$
4. Constraints $2x + y + z + r = 3$ and $x + y + r + s = 4$
5. Constraints $x + 2y + r = 3$, $3x - y + 3z + s = 5$, $4y + t = 1$

If a canonical lpp fails to be perfect because the objective function depends on at least one basic variable, then the objective function can be rewritten so as to obtain an equivalent perfect canonical lpp. In fact, if a basic variable, say r, occurs in the objective function, then, in the constraint containing r, we solve for r to obtain an expression involving the other (non-basic) variables in that constraint, and we substitute this expression for r in the objective function. If we do this for all such basic variables, the new form of the objective function depends only on non-basic variables, and the resulting canonical lpp is perfect.

Example 3.5

Consider the following canonical lpp.

$$\text{Maximize } u = 3x - y + 2z - 3s$$

subject to the constraints

$$4x - 3y - z + s = 2$$
$$x + y + 2z + t = 3$$

The basic variables are s and t, and the basic point $(0,0,0,2,3)$ is feasible. However, the objective function depends on the basic variable s. Solving the constraint containing s for s, we obtain $s = 2 - 4x + 3y + z$. Hence, the objective function u is equal to $3x - y + 2z - 3(2 - 4x + 3y + z) = 15x - 10y - z - 6$, which depends only on the non-basic variables x, y, and z. With this new form of the objective function, we have an equivalent perfect canonical lpp.

Example 3.6

Consider the following standard maximization problem:

$$\text{Maximize } u = 3x + 5y$$

subject to the constraints

$$x + y \leq 6$$
$$2x + y \leq 8$$

See the diagram in Figure 3.1.

The feasible set is the interior and boundary of quadrilateral OABC. By the Vertex Theorem, the maximum value of u on the constraint set occurs at one of the vertices $O(0,0)$, $A(4,0)$, $B(2,4)$, $C(0,6)$. By simple calculation, the maximum $u = 30$ occurs at $C(0,6)$. Now consider the following alternative approach. An equivalent slack system is:

$$\text{Maximize } u = 3x + 5y$$

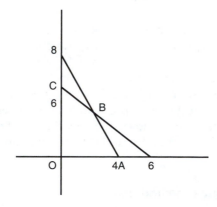

Figure 3.1

subject to the constraints

$$x + y + r = 6$$

$$2x + y + s = 8$$

The basic point is $(0,0,6,8)$. The corresponding values $x = 0$ and $y = 0$ yield the value $u = 0$, which clearly is not the maximum. Let us find an equivalent canonical lpp with y and s as basic variables. Solving the first constraint for y, we get $y = 6 - r - x$, and substituting this for y in the second constraint, we obtain $2x + 6 - r - x + s = 8$, which reduces to $x - r + s = 2$. Thus, we have the new constraints

$$x + y + r = 6$$

$$x - r + s = 2$$

The new basic point is $(0,6,0,2)$, with the corresponding point $x = 0$, $y = 6$, that is, the point C. The new objective function is $u = 3x + 5(6 - r - x) = 30 - 5r - 2x$. We now have an equivalent perfect canonical lpp. Since r and x are non-negative, we see that 30 is the maximum value of u and it is attained when $x = 0$ and $r = 0$, that is, at the new basic point.

Of course, we were lucky to have chosen y and s as new basic variables. The choice of a different pair, say x and s, would not have led us to the maximum. (Verify that one would get the non-feasible basic point $(6,0,0,-4)$ and that u would become $18 + 2y - 3r$.) So, it would be nice to find a method that, in cases where the given lpp is solvable, eventually yields an equivalent perfect canonical lpp for which the objective function must have a maximum value at the basic point. In cases where the given lpp has no solution, the method should tell us so. Such a method, the Simplex Method, does exist. We shall now proceed to its description, justification, and application.

3.2   The simplex method

If a canonical lpp is given, we can transform each constraint equation into an equation in which the right side of the equation is the negative of the basic variable of the constraint. The resulting lpp is called a *standard canonical lpp*. The simplex method will be formulated in terms of such lpp's.

Example 3.7
Consider the standard maximization lpp:

$$\text{Maximize } u = 3x + 5y - 2z$$

subject to the constraints

$$x - 4y + z \le -2$$

$$2x + 3y - 4z \le 5$$

Introduce slack variables r and s to obtain a canonical lpp:

$$x - 4y + z + r = -2$$

$$2x + 3y - 4z + s = 5$$

The basic variables are r and s. Now we can obtain the following equivalent standard canonical lpp:

$$\text{Maximize } u = 3x + 5y - 2z$$

subject to the constraints

$$x - 4y + z + 2 = -r$$

$$2x + 3y - 4z - 5 = -s$$

Exercise 3.3
Find a standard canonical lpp equivalent to the following lpp:

$$\text{Maximize } u = 12x - 4y$$

with constraints

$$3x - y \leq 2 \quad \text{and} \quad 2x + 3y \leq 4$$

A standard canonical lpp will be written in a certain abbreviated form, called a *simplex tableau*. We shall illustrate this in terms of two specific examples, and then we shall give the general definition.

Example 3.8
Consider the following standard canonical lpp:

$$\text{Maximize } u = 2x + 3y - 2z + 7$$

subject to the constraints

$$4x - 5y + z - 2 = -r$$

$$x + 2y - 4z - 5 = -s$$

The corresponding simplex tableau is:

x	y	z	1	
4	−5	1	−2	$= -r$
1	2	−4	−5	$= -s$
2	3	−2	7	$= u$

The top row always consists of the non-basic variables and the number 1. The second row represents the first constraint. Each of the entries 4, –5, 1, –2 is multiplied by the corresponding entry in the top row, and the results are added and set equal to –r. Similarly, the third row represents the second constraint. The last row represents the objective function $u = 2x + 3y - 2z + 7$. The last row is called the *objective row*. The column of numbers under "1" and above the objective row is called the *column of constants*. In this example, it consists of –2 and –5. Note that the basic point $(0,0,0,2,5)$ is feasible, so that we are dealing with a perfect canonical lpp. This is always so when the column of constants consists of non-positive numbers, as it did in this example. Observe also that the entry in the lower right-hand corner is the value of the objective function at the basic point. In this example, that entry is 7, the value of $u = 2x + 3y - 2z + 7$ when $x = y = z = 0$.

Example 3.9
The standard canonical lpp:

$$\text{Maximize } u = 5r + 2y$$

subject to the constraints

$$-3r + 4y - 2 = -x$$
$$r - 3y - 1 = -s$$
$$r + 2y - 4 = -t$$

yields the simplex tableau

r	y	1	
-3	4	-2	$= -x$
1	-3	-1	$= -s$
1	2	-4	$= -t$
5	2	0	$= u$

The basic point $(2,0,0,1,4)$ is feasible. (The variables are taken in the order x, y, r, s, t.) This canonical lpp is perfect.

In general, consider a standard canonical lpp with objective function $u = c_1 x_1 + \cdots + c_n x_n + d$ and m constraints

$$a_{11}x_1 + a_{12}x_2 + \cdots + a_{1n}x_n - b_1 = -x_{n+1}$$
$$a_{21}x_1 + a_{22}x_2 + \cdots + a_{2n}x_n - b_2 = -x_{n+2}$$

$$\cdots\cdots\cdots\cdots$$

$$a_{m1}x_1 + a_{m2}x_2 + \cdots + a_{mn}x_n - b_m = -x_{n+m}$$

The non-basic variables are $x_1, \ldots, x_n$ and the basic variables are $x_{n+1}, \ldots, x_{n+m}$. The corresponding simplex tableau is:

x_1	$\ldots$	x_n	1	
a_{11}	$\ldots$	a_{1n}	$-b_1$	$= -x_{n+1}$
a_{21}	$\ldots$	a_{2n}	$-b_2$	$= -x_{n+2}$
$\ldots$	$\ldots$	$\ldots$	$\ldots$	$\ldots$
a_{m1}	$\ldots$	a_{mn}	$-b_m$	$= -x_{n+m}$
c_1	$\ldots$	c_n	d	$= u$

We wish to use simplex tableaux to get closer to and finally reach a solution of a lpp, if such a solution exists. In the course of this procedure, if we have a simplex tableau that does not immediately yield a solution, we shall solve one of the constraints for a non-basic variable y and then replace y in the other constraints and in the objective function. Then y becomes a basic variable; at the same time, the variable that was the basic variable of the constraint that was solved for y becomes a non-basic variable. We shall give precise instructions for doing all this, but first let us look at an example.

Example 3.10

Let us look at the standard canonical lpp of Example 3.9. The form $u = 5r + 2y$ of the objective function does not reveal how to find the maximum value of u subject to the given constraints. We shall use the first constraint to replace the basic variable x by y. (Rules for choosing the appropriate constraint and the new basic variable will be given later.)

Solving for $-y$, we get $-y = \tfrac{1}{4} x - \tfrac{3}{4} r - \tfrac{1}{2}$. We must then replace y in the other two constraints and in the objective function:

$$r + 3(\tfrac{1}{4} x - \tfrac{3}{4} r - \tfrac{1}{2}) - 1 = -s$$

$$r - 2(\tfrac{1}{4} x - \tfrac{3}{4} r - \tfrac{1}{2}) - 4 = -t$$

$$u = 5r - 2(\tfrac{1}{4} x - \tfrac{3}{4} r - \tfrac{1}{2})$$

Simplifying, we get the constraints

$$\tfrac{1}{4} x - \tfrac{3}{4} r - \tfrac{1}{2} = -y$$

$$\tfrac{3}{4} x - \tfrac{5}{4} r - \tfrac{5}{2} = -s$$

$$-\tfrac{1}{2} x + \tfrac{5}{2} r - 3 = -t$$

and objective function $u = -\tfrac{1}{2} x + \tfrac{13}{2} r + 1$.

The corresponding simplex tableau is:

x	r	1	
$\tfrac{1}{4}$	$-\tfrac{3}{4}$	$-\tfrac{1}{2}$	$= -y$
$\tfrac{3}{4}$	$-\tfrac{5}{4}$	$-\tfrac{5}{2}$	$= -s$
$-\tfrac{1}{2}$	$\tfrac{5}{2}$	-3	$= -t$
$-\tfrac{1}{2}$	$\tfrac{13}{2}$	1	$= u$

The value of the objective function at the basic point is always the entry in the lower right-hand corner of the tableau. In this example, the value of $u = -\frac{1}{2}x + \frac{13}{2}r + 1$ at the basic point (where $x = 0$ and $r = 0$) is 1, the entry in the lower right-hand corner of the tableau. Note also that, since the column of constants is all non-positive, we still have a perfect canonical lpp.

Moreover, the new simplex tableau is an improvement over the previous one, since the positive coefficient 5 under r in the previous objective row has been replaced by the negative coefficient $-\frac{1}{2}$ of x. The reason that this is an improvement is that we eventually want to obtain a simplex tableau in which the coefficients of variables in the objective row (that is, all the numbers in the objective row except the last number, under the column of constants) are non-positive. For, in that case, the objective function u has the form $c_1 y_1 + \cdots + c_n y_n + d$, where each $c_i \leq 0$. Therefore, u attains its maximum value d when every y_i is 0, and the lpp is solved. (The values of the basic variables are obtained as the negatives of the entries in the column of constants.)

3.3   Pivoting

In the course of applying the simplex method, we choose one of the constraints, say, $a_{i1}x_1 + a_{i2}x_2 + \cdots + a_{in}x_n - b_i = -x_{n+i}$ and we decide to change one of the non-basic variables, say x_j, into a basic variable, and to change the basic variable x_{n+i} into a non-basic variable. We do this by solving the constraint equation for $-x_j$ and using the result to replace x_j in the other constraint equations and in the objective function. (Of course, to solve the constraint equation for $-x_j$, the coefficient a_{ij} cannot be 0.) The new canonical lpp and the new simplex tableau are said to be obtained by *pivoting* with respect to a_{ij}, and a_{ij} is called the *pivot*, the ith row is called the *pivot row* and the jth column is called the *pivot column*.

When the number of variables and constraints is small, the pivoting can be carried out by hand. This was the case in Examples 3.9 and 3.10, where we pivoted with respect to $a_{12} = 4$. But, when the number of variables and constraints is large or when we wish to prove something about pivoting, we need to derive formulas for the new tableau. So, let us assume we start with the tableau

x_1	.....	x_n	1	
a_{11}	.....	a_{1n}	$-b_1$	$= -x_{n+1}$
a_{21}	.....	a_{2n}	$-b_2$	$= -x_{n+2}$
...	.....	...	...	......
a_{m1}	.....	a_{mn}	$-b_m$	$= -x_{n+m}$
c_1	.....	c_n	d	$= u$

and that we pivot with respect to a_{ij}. We first solve the ith constraint equation

$$a_{i1}x_1 + a_{i2}x_2 + \cdots + a_{ij}x_j + \cdots + a_{in}x_n - b_i = -x_{n+i}$$

for $-x_j$, obtaining the new ith constraint equation

$$(a_{i1}/a_{ij})x_1 + (a_{i2}/a_{ij})x_2 + \cdots + (1/a_{ij})x_{n+i} + \cdots + (a_{in}/a_{ij})x_n - (b_i/a_{ij}) = -x_j$$

Now look at the kth constraint, with $k \neq i$. When we replace x_j and do some elementary calculation, the new kth constraint will have the form

$$a_{k1}{}^* x_1 + a_{k2}{}^* x_2 + \cdots + cx_{n+i} + \cdots + a_{kn}{}^* x_n - b_k{}^* = -x_{n+k}$$

where, for $p \neq j$, $a_{kp}{}^* = a_{kp} - (a_{kj}a_{ip}/a_{ij})$, and $c = -a_{kj}/a_{ij}$, and $b_k{}^* = (b_i a_{kj}/a_{ij}) - b_k$. The new form of the objective function will be

$$c_1{}^* x_1 + c_2{}^* x_2 + \cdots + Dx_{n+i} + \cdots + c_n{}^* x_n{}^* + d^*,$$

where, for $q \neq j$, $c_q{}^* = c_q - (c_j a_{iq}/a_{ij})$, and $D = -c_j/a_{ij}$, and $d^* = d + (b_i/a_{ij})$.

3.4 The perfect phase of the simplex method

We shall explain how the simplex method applies to a perfect standard canonical lpp, that is, we start with a simplex tableau

x_1	.....	x_n	1	
a_{11}	.....	a_{1n}	$-b_1$	$= -x_{n+1}$
a_{21}	.....	a_{2n}	$-b_2$	$= -x_{n+2}$
...	.....	...	...	......
a_{m1}	.....	a_{mn}	$-b_m$	$= -x_{n+m}$
c_1	.....	c_n	d	$= u$

in which every b_i is non-negative.

In the course of describing the simplex method, we sometimes will need to know certain facts whose proofs would distract us from the main line of our argument. In such cases, we will just take note of those facts and postpone their proofs until later. We already have shown that, if every c_j is negative or zero, then we have a solution. Hence, we may assume that at least one c_j is positive.

Step 1.   Choose some positive c_j.

Fact 1.   If all the entries a_{ij} in the column above c_j are negative or zero, then the lpp has no solution.
Hence, let us assume that at least one a_{ij} is positive. For every such a_{ij}, calculate the ratio b_i/a_{ij}.

Step 2.   Pivot with respect to that positive a_{ij} for which the ratio b_i/a_{ij} is minimal.

Fact 2
1.   The result is a perfect standard canonical lpp, that is, the numbers in the column of constants in the resulting tableau are negative or zero.

2. The value of the objective function at the new basic point is greater than or equal to its value at the previous basic point. Let the *i*th row be the pivot row. If b_i is positive, the value of the objective function is strictly bigger than before. When $b_i = 0$ (this is called the *degenerate case*), then the value of the objective function does not change. Moreover, in the degenerate case, the entire column of constants remains the same, and the new basic point is the same as the previous basic point.

Example 3.11

Consider the tableau of Example 3.9. Both entries 5 and 2 of the objective row are positive. As Step 1, choose 2. In the column above this 2, there are two positive entries 4 and 2, with corresponding ratios 2/4 and 4/2. Since 2/4 < 4/2, in Step 2 we pivot with respect to 4. This is what we did in Example 3.10, with the resulting tableau shown in that example. Note that the numbers $-1/2$, $-5/2$, and -3 in the column of constants of that tableau are all negative, so that the resulting lpp is perfect. Since the pivot row was the first row and $b_1 = 2 > 0$, we should expect, by virtue of Fact 2(2), that the objective function would increase at the basic point. In fact, it increased from $u = 5r + 2y = 0$ at the basic point where $r = 0$, $y = 0$, to $u = -1/2 x + \frac{13}{2} r + 1 = 1$ at the basic point where $x = 0$, $r = 0$.

Exercise 3.4

In Example 3.11, in Step 1 choose entry 5 instead of 2 in the objective row and construct the resulting tableau. What solution do you eventually obtain?

Step 3. Keep on repeating Steps 1 and 2.

There are three possible outcomes.

(A) We eventually reach a tableau in which all the numbers in the objective row are negative. As we have seen, this yields a solution of the lpp.

(B) We eventually reach a tableau to which Fact 1 is applicable and then we know that the lpp has no solution.

(C) We keep on repeating Steps 1 and 2 forever, that is, cases (A) and (B) do not occur.

Fact 3. There is a way of modifying the Simplex Method so that case (C) does not occur.

Example 3.12

Let us continue with the problem of Example 3.9. We have reached the tableau in Example 3.10. Since 13/2 is the only positive coefficient c_i in the objective row, we look at the numbers in the column above 13/2. Since 5/2 is the only one that is positive, we pivot with respect to 5/2. The resulting tableau is

x	t	1	
1/10	3/10	−7/5	$= -y$
1/2	1/2	−4	$= -s$
−1/5	2/5	−6/5	$= -r$
4/5	− 13/5	44/5	$= u$

Since 4/5 is the only positive coefficient in the objective row, we look at the positive numbers 1/10 and $^1/_2$ in the column above 4/5. Since $4/(1/2) <$ $(7/5)/(1/10)$, we pivot with respect to $^1/_2$. The resulting tableau is

s	t	1	
−1/5	1/5	−3/5	$= -y$
2	1	−8	$= -x$
2/5	3/5	−14/5	$= -r$
−8/5	−17/5	76/5	$= u$

We have reached case (A), which yields a solution. The maximum value of $5r + 2y$ is 76/5, which is attained at $r = 14/5$ and $y = 3/5$.

Example 3.13

Let us consider the lpp of Example 3.4. Taking r and s as basic variables, we get the tableau

x	y	z	1	
3	1	−1	−4	$= -r$
2	0	−3	−7	$= -s$
4	−2	3	0	$= u$

Since this is perfect, we can apply the simplex method. There are two positive coefficients in the objective row, but the column over 3 has only negative entries. Hence, we look at the column over 4. Since $4/3 < 7/2$, we pivot with respect to 3 in the first row, first column. The resulting tableau is

r	y	z	1	
1/3	1/3	−1/3	−4/3	$= -x$
−2/3	−2/3	−7/3	−13/3	$= -s$
−4/3	−10/3	13/3	16/3	$= u$

Since all the entries in the column above the positive coefficient 13/3 in the objective row are negative, case (B) applies and our lpp has no solution.

Exercise 3.5

Apply the Simplex Method to the following lpp's.

1. The lpp in Example 3.8.
2. Maximize $u = -5x + 3y$, subject to $x + 2y \le 3$ and $3x - y \le 4$.
3. Maximize $u = x - y - z$, subject to $x - y + z \le 4$ and $x + y - z \le 2$.

4. Maximize $u = 3x - 3y - 2z + 4$, subject to $-5x + 4y + z \le 10$ and $-4x - 2y + 2z \le 9$.

5. Maximize $u = 3x + 4y - 2z + 2w + 5$, subject to $x + 2y - z + 3w \le 1$ and $2x - y + 4z + w \le 2$.

6. Maximize $u = -3x - y + 2z - 12$, subject to $-4x - 2y + z \le 5$ and $3x + y \le 2$.

7. Maximize $u = x - 2y - 4z + 8$, subject to $x + 2y + 8z \le 5$, $-7x + 2y + 6z \le 2$, and $2x + y + 3z \le 6$.

8. Maximize $u = x + 2y$, subject to $4x + 3y \le 12$ and $-x + y \le 2$.

9. Maximize $u = 4x + 7y + 2z$, subject to $3x - y + 5z \le 7$ and $x + 2y - z \le 3$.

Exercise 3.6
If $A = (a_1, \ldots, a_n)$ and $B = (b_1, \ldots, b_n)$ are solutions of a standard maximization lpp, show that all the vectors of the form $tA + (1-t)B$, for $0 \le t \le 1$, are solutions also. (Those vectors make up the "segment" connecting points A and B.)

Example 3.14 (Multiple solutions)
Consider the standard maximization lpp:

Maximize $u = 2x + 2y + 22$, subject to $x + y \le 5$ and $-x + 3y \le 7$. This gives rise to the tableau

	x	y	1	
	1	1	-5	$= -r$
	-1	3	-7	$= -s$
	2	2	22	$= u$

Pivoting on $a_{11} = 1$ yields

	r	y	1	
	1	1	-5	$= -x$
	1	4	-12	$= -s$
	-2	0	32	$= u$

Thus, the maximum is 32 at $x = 5$, $y = 0$, which is point A in the diagram below.

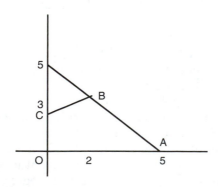

The constraint set is the solid quadrilateral OABC. Now, when a maximum is reached at a tableau having a zero in the objective row, it may happen that the maximum also can be attained at a different point. We treat that zero as we would a positive entry in the objective row. This leads us to pivot with respect to $a_{22} = 4$, yielding the tableau

r	s	1	
3/4	−1/4	−2	$= -x$
1/4	1/4	−3	$= -y$
−2	0	32	$= u$

Note that the objective row is unchanged, but now the maximum 32 is reached at $x = 2$ and $y = 3$, the vertex B of the quadrilateral. By Exercise 3.6, all points on the segment AB are solutions of the lpp. (Observe that the segment AB is parallel to the level lines $2x + 2y + 22 = c$ of the function u.)

Exercise 3.7
Find multiple solutions (if any) of the following lpp's.

1. Maximize $u = -3x - 2z + 7$, subject to constraints $-x + y + 4z \le 2$ and $2x + y + z \le 1$.
2. Maximize $u = 6x + 4y - 3$, subject to constraints $3x + 2y \le 6$ and $7x + 2y \le 7$.

3.5   The Big M method

Our simplex method presupposes that we start with a perfect standard canonical lpp. If that is not so, then we must reduce the lpp to a perfect standard canonical lpp.

Example 3.15
Consider the following standard maximization lpp.

Maximize $u = 3x - 7y + 5z$, subject to the constraints $2x + 5y - z \le 3$, $4x - y + 3z \le -2$ and $x + 2y - 4z \le -3$.

Introducing slack variables r, s, and t, we get constraints $2x + 5y - z + r = 3$, $4x + 5y - z + s = -2$, and $x + 2y - 4z + t = -3$. This is not perfect because the constants −2 and −3 are negative.

There are several ways to produce a suitable perfect lpp whose solution is equivalent to the solution of a given lpp, and we will use the so-called *Big M* method. Let us assume that we already have introduced slack variables. The problem that can occur now is that at least one constant on the right side of a constraint equation might be negative. In such a case, we multiply both sides of the equation by –1. Now, all constants on the right side of the equations are non-negative. However, the new lpp may no longer be canonical because at least one equation may fail to have a distinguished variable with coefficient 1.

Example 3.15 (Continued)
The constraints equations are now:

$$2x + 5y - z + r = 3,$$

$$-4x - 5y + z - s = 2, \quad \text{and}$$

$$-x - 2y + 4z - t = 3$$

The second equation fails to have a suitable distinguished variable. The coefficient of s is –1, not 1, and the variable z cannot serve as a distinguished variable because it occurs in another equation. A similar difficulty holds for the third equation.

For each constraint equation that now lacks a distinguished variable, add a new "artificial" variable as a term on the left side. For each such variable A, we subtract a term MA from the objective function, where M is taken to be a very large positive constant. We now have a perfect canonical lpp. (The objective function can be represented in terms of the original variables by solving the constraints for the artificial variables and replacing the latter in the objective function.)

Example 3.15 (Continued)
The constraint equations become:

$$2x + 5y - z + r = 3,$$

$$-4x - 5y + z - s + A = 2, \quad \text{and}$$

$$-x - 2y + 4z - t + B = 3$$

where A and B are the new artificial variables. The new objective function is

$$3x - 7y + 5z - MA - MB$$

Observe that we now have a perfect canonical lpp, since the new artificial variables A and B serve as distinguished variables in their constraint equations. In fact, from the constraints we get $A = 4x + 5y - z + s + 2$ and $B = x + 2y - 4z + t - B + 3$, so that the objective function is $3x - 7y + 5z - MA - MB = 3x - 7y + 5z - M(4x + 5y - z + s + 2) - M(x + 2y - 4z + t - B + 3) = (3 - 5M)x - (7 + 7M)y + (5 + 5M)z - Ms - Mt - 5M$.

Apply the simplex method to the resulting perfect canonical lpp.

Fact 4. In the Big M method, if there is a solution, then all the artificial variables have the value zero and we obtain a solution of the original lpp.

Example 3.15   (Continued)
The new objective function is

$$(3 - 5M)x - (7 + 7M)y + (5 + 5M)z - Ms - Mt - 5M$$

and the constraints can be written as:

$$2x + 5y - z - 3 = -r$$

$$-4x - 5y + z - s - 2 = -A$$

$$-x - 2y + 4z - t - 3 = -B$$

and we get the tableau

x	y	z	s	t	1	
2	5	−1	0	0	−3	$= -r$
−4	−5	1	−1	0	−2	$= -A$
−1	−2	4	0	−1	−3	$= -B$
$3 - 5M$	$-7 - 7M$	$5 + 5M$	$-M$	$-M$	$-5M$	$= u$

Since M can be chosen as large as we want, the entries $3 - 5M$, $-7 - 7M$, $-M$, and $-M$ in the objective row can be assumed to be negative, so that the only positive entry in the objective row is $5 + 5M$. Examining the ratios $2/1$ and $^3/_4$ for the positive entries 1 and 4 in the column above $5 + 5M$, we see that we must pivot with respect to $a_{31} = 4$. The resulting tableau is:

x	y	B	s	t	1	
7/4	9/2	1/4	0	−1/4	−15/4	$= -r$
−15/4	−9/2	−1/4	−1	0	−5/4	$= -A$
−1/4	−1/2	1/4	0	−1/4	−3/4	$= -z$
$17/4 - 15M/4$	$-(9/2 + 9M/2)$	$-(5 + 5M)/4$	$-M$	$(5 + M)/4$	$(15 - 5M)/4$	$= u$

The entries in the column above the positive entry $(5 + M)/4$ in the objective row are negative or zero. Hence, the lpp has no solution, by virtue of Fact 1. Therefore, the original lpp has no solution.

Example 3.16
Consider the following lpp.

Maximize $u = 5x - 3y$, subject to $3x + 7y \leq 5$ and $2x - 3y \leq -2$. Introducing slack variables, we get constraints

$$3x + 7y + r = 5 \quad \text{and}$$

$$2x - 3y + s = -2$$

Multiplying the second constraint by −1 yields

$$3x + 7y + r = 5 \quad \text{and}$$

$$-2x + 3y - s = 2$$

So, we must introduce an artificial variable:

$$3x + 7y + r = 5 \quad \text{and}$$

$$-2x + 3y - s + A = 2$$

and the new objective function is $5x - 3y - MA$. Solving the constraint for A and substituting in the objective function yields

$$5x - 3y - M(2x - 3y + s + 2) = (5 - 2M)x - 3(1 - M)y - Ms - 2M$$

The corresponding tableau is:

	x	y	s	1	
	3	7	0	−5	$= -r$
	−2	3	−1	−2	$= -A$
	$5 - 2M$	$3M - 3$	$-M$	$-2M$	$= u$

The only positive entry in the objective row is $3M - 3$. Then we pivot with respect to $a_{22} = 3$ and obtain the new tableau

	x	A	s	1	
	23/3	−7/3	7/3	−1/3	$= -r$
	−2/3	1/3	−1/3	−2/3	$= -y$
	3	$1 - M$	−1	−2	$= u$

Next we pivot with respect to $a_{11} = 23/3$, obtaining the tableau

	r	A	s	1	
	3/23	−7/23	7/23	−1/23	$= -x$
	2/23	3/23	−3/23	−16/23	$= -y$
	−9/23	$(44/23) - M$	−44/23	−43/23	$= u$

All the entries in the objective row are negative, since $44/23 - M$ is negative for sufficiently large M. Hence, a solution has been obtained. The maximum is $-43/23$ at $x = 1/23$ and $y = 16/23$. (The reader should draw a graph of the constraint set.) Notice that the solution assigns zero to the artificial variable A.

Exercise 3.8
Apply the Big M method to the following lpp's:

1. Maximize $u = x + y$, subject to $x - 2y \le -4$ and $3x - y \le -2$.
2. Maximize $u = x - 2y + 3z$, subject to $3x + y - z \le -1$, $x - y + 2z \le 3$, and $2x + 3y - 3z \le -2$.

3. Maximize $u = x + 4y$, subject to $2x - y \leq -6$, $x - y \leq -7$, and $x + y \leq 5$.
4. Minimize $u = -x + 9y + z$, subject to $x + y - z \geq 7$ and $-3x + 4y + 2z \geq 3$. (Hint: Transform into a maximization problem.)

3.6 Bland's rules to prevent cycling[2]

Our present rules for the simplex method very occasionally result in an infinite cycle of pivots, as in the following example.

Example 3.17

$$\text{Maximize } u = 4x - (7/2)y - 33z - 3w + 3$$

subject to

$$-32x - 4y + 36z + 4w \leq 0,$$

$$4x + (3/2)y - 15z - 2w \leq 0,$$

$$y \leq 1$$

Introducing slack variables r, s, and t, we get the tableau

x	y	z	w	1	
−32	−4	36	4	0	$= -r$
4	3/2	−15	−2	0	$= -s$
0	1	0	0	−1	$= -t$
4	−7/2	−33	−3	3	$= u$

We must pivot with respect to $a_{12} = 4$, getting the new tableau

s	y	z	w	1	
8	8	−84	−12	0	$= -r$
1/4	3/8	−15/4	−1/2	0	$= -x$
0	1	0	0	−1	$= -t$
−1	2	−18	−1	3	$= u$

For our next pivot, we have a choice between 8 and 3/8 in the second column, and we arbitrarily choose 8. This produces the tableau

s	r	z	w	1	
1	1/8	−21/2	−3/2	0	$= -y$
−1/8	−3/64	3/16	1/16	0	$= -x$
−1	−1/8	21/2	3/2	−1	$= -t$
−3	−71/4	3	2	3	$= u$

[2]Here, as in much of this chapter, our exposition relies a great deal on Sultan [1993].

We select the third column for the next pivot, which must then take place with respect to $a_{13} = 3/16$. This yields the tableau

s	r	x	w	1	
−6	−5/2	56	2	0	= −y
−2/3	−1/4	16/3	1/3	0	= −z
6	5/2	−56	−2	−1	= −t
−1	1/2	−16	1	3	= u

Next we decide to pivot in the fourth column, and we choose $a_{14} = 2$ as the pivot. The resulting tableau is

s	r	x	y	1	
−3	−5/4	28	1/2	0	= −w
1/3	1/6	−4	−1/6	0	= −z
0	0	0	1	−1	= −t
2	−7/4	−44	−1/2	3	= u

The next pivot must be with respect to $a_{21} = 1/3$, with the resulting tableau

z	r	x	y	1	
9	1/4	−8	−1	0	= −w
3	1/2	−12	−1/2	0	= −s
0	0	0	1	−1	= −t
−6	3/4	−20	1/2	3	= u

Our next pivot is chosen to be $a_{12} = {}^1\!/_4$. The new tableau is

z	w	x	y	1	
36	4	−32	−4	0	= −r
−15	−2	4	3/2	0	= −s
0	0	0	1	−1	= −t
−33	−3	4	−7/2	3	= u

Notice that this tableau is the same as the original tableau, except that the columns have been permuted. We can then repeat the pivots that were made and come back to essentially the same tableau again. Since this can go on forever, the simplex method does not yield a result in this case. Hence, we must modify the simplex method in order to avoid such cycles. An appropriate modification was given in Bland [1977] and will be explained below.

First note that a canonical lpp has only a finite number of basic points. If there are k non-basic variables and j basic variables, then a basic point determines a choice of k of the $k + j$ variables and different basic points determine different choices. Hence there are at most $\frac{(k+j)!}{k!\,j!}$ basic points, since $\frac{(k+j)!}{k!\,j!}$ is the number of ways of choosing k objects from a set of $k+j$ objects.

Recall from Fact 2(2) that there are two cases in every pivot. If the pivot row is the ith row and $-b_i$ is the number that appears in the column of constants in the ith row, then the non-degenerate case occurs when $b_i > 0$, and the degenerate case occurs when $b_i = 0$.

In the non-degenerate case, the value of the objective function at the new basic point is greater than its value at the old basic point, whereas, in the degenerate case, the value of the objective function at the new basic point is the same as the value of the objective function at the old basic point. In particular, after every non-degenerate pivot operation, the new basic point is different from all of the preceding basic points.

Therefore, since any canonical lpp has only a finite number of possible basic points, any infinite cycle of tableaux could contain only a finite number of non-degenerate pivots and so, eventually all the pivots would have to be degenerate. Bland's Rules will prevent this from happening.

Bland's rules for choosing a pivot

We shall assume a specific ordering of the variables. For convenience, we designate the variables by $x_1, x_2, \ldots, x_n$. The first variable in the ordering is x_1, the second one is x_2, and so on. Let us assume that we are given a tableau and we must choose a pivot.

Rule 1. If there are several positive entries in the objective row, we choose the entry for which the variable at the top of its column has the least subscript.

Rule 2. If Rule 1 has been applied and the chosen column, say the kth, contains several entries a_{ik} for which the corresponding ratios b_i/a_{ik} are minimal, then we choose as pivot element the one for which the variable in the right-most column has the least subscript.

Example 3.18

(a) Consider the tableau

x_4	x_1	x_5	1	
3	0	8	0	$= -x_3$
6	5	−1	0	$= -x_2$
2	−6	4	9	$= u$

The objective row contains two positive entries, 2 and 4. The variables heading the columns above 2 and 4 are x_4 and x_5. Since the subscript 4 is less than the subscript 5, Bland's Rule 1 tells us to choose the entry 2 in the objective row. The two entries 3 and 6 in the column above 2 have corresponding equal ratios 0/3 and 0/6. We look at the right-most variables x_3 and x_2 in those rows and, by Bland's Rule 2, we choose the variable x_2 with the smaller subscript. Therefore, we pivot with respect to $a_{21} = 6$.

(b) Consider the tableau

x_3	x_2	1	
0	2	0	$= -x_5$
−1	1	0	$= -x_1$
3	6	5	$= -x_4$
4	5	2	$= u$

Of the two positive entries 4 and 5 in the objective row, Bland's Rule 1 requires us to select 5, since the variable x_2 in the column above 5 has a smaller subscript than the variable x_3 in the column above 4. Next, in the column above 5, there are three positive entries 2, 1, and 6, with corresponding ratios 0/2, 0/1, and 5/6. Since the entries 2 and 1 both have the minimal corresponding ratio 0, we look at the variables x_5 and x_1 in the rows containing 2 and 1. Bland's Rule 2 tells us to choose the row with variable x_1, since the subscript 1 is smaller than the subscript 5. Hence, the required pivot element is $a_{22} = 1$.

Example 3.19

Let us see how Bland's Rules deal with the lpp that gave rise to a cycle in Example 3.17. We shall order the variables as follows: x, y, z, w, r, s, t (in the role of $x_1, x_2, x_3, x_4, x_5, x_6, x_7$ in the general treatment above). The first difference from our earlier work occurs when we reach the tableau

s	y	z	w	1	
8	8	−84	−12	0	$= -r$
1/4	3/8	−15/4	−1/2	0	$= -x$
0	1	0	0	−1	$= -t$
−1	2	−18	−1	3	$= u$

Here, Bland's Rules dictate that we pivot with respect to 3/8 (rather than 8 as before). This yields the tableau

s	x	z	w	1	
8/3	64/3	−4	−4/3	0	$= -r$
2/3	8/3	−10	−4/3	0	$= -y$
−2/3	−8/3	10	4/3	−1	$= -t$
−7/3	−16/3	2	5/3	3	$= u$

Bland's Rules then tell us to pivot with respect to $a_{33} = 10$, resulting in the tableau

s	x	t	w	1	
12/5	304/15	−2/5	−4/5	−2/5	$= -r$
0	0	1	0	−1	$= -y$
−1/15	−4/15	1/10	2/15	−1/10	$= -z$
−11/5	−24/5	−1/5	7/5	16/5	$= u$

The next pivot must be with respect to $a_{34} = 2/15$, yielding the tableau

s	x	t	z	1	
2	56/3	1/5	6	−1	$= -r$
0	0	1	0	−1	$= -y$
−1/2	−2	3/4	15/2	−3/4	$= -w$
−3/2	−2	−5/4	−21/2	17/4	$= u$

Hence, the maximum is 17/4, attained at $x = z = 0$, $y = 1$, $w = {}^3/_4$.

Exercise 3.9

For the following tableaux, find the pivot element required by Bland's Rules.

(a)

x_1	x_2	x_3	x_4	1	
−32	−4	36	4	1	$= -x_5$
4	3/2	−15	−2	0	$= -x_6$
0	1	0	0	0	$= -x_7$
−2	3	−33	1	3	$= u$

(b)

x_4	x_2	1	
0	1	2	$= -x_1$
1	5	0	$= -x_5$
6	3	0	$= -x_3$
4	2	1	$= u$

Let us show now that Bland's Rules prevent the occurrence of infinite cycles of degenerate pivots. So, let us assume that, contrary to what we want to prove, there is an infinite cycle of degenerate pivots. To make the reasoning clearer, we shall work with a specific case, but the argument generalizes to all cases. Let us assume that $y_1, \ldots, y_{15}$ are the variables in increasing order that occur infinitely often as basic variables in the sequence of degenerate pivots. (For example, $y_1, \ldots, y_{15}$ might be $x_3, x_7, x_9, \ldots, x_{34}$.)

For our further convenience, we assume that our tableaux are constructed so that the non-basic variables at the head of the columns occur in order from left to right and that the basic variables of the cycle in the right-most column are listed first in order from top to bottom. Let us consider a tableau that occurs well along in the sequence just one pivot before the variable y_{15} becomes basic.

x_i preceding y_{15} y_{15} x_i after y_{15} 1.

− − − − − − − − − − − − +		d

In the objective row, the entry in the column under y_{15} must be positive, but all the preceding entries in the objective row must be non-positive. After the pivot, y_{15} is basic, and, since y_{15} is the greatest variable occurring in the cycle, all non-basic variables x_i after y_{15} remain non-basic thereafter. Now let us look at the tableau just one pivot before y_{15} becomes non-basic again. The variable that becomes basic after that pivot precedes y_{15}. For definiteness, let us assume it is y_8.

y_8 1.

(−)	0	$= -y_{j_2}$
(−)	0	$= -y_{j_2}$
(−)	.....	.....
(+)	0	$= -y_{15}$
....		
c_8	d	

By Bland's Rules and our convention about the arrangement of the rows, the entries in the column under y_8, starting from the top row, are all non-positive until the row containing y_{15}, where the entry is positive. In addition, the entry c_8 is positive, since the pivot took place in the column under y_8.

Let us define a point P in the following way. Let $y_8 = 0$ and set all the other non-basic variables in this second tableau equal to zero. Then, at point P, all basic variables in this tableau above y_{15} receive a non-positive value, since the entries in the column under y_8 and in the corresponding rows are all non-positive and the entries in those rows in the column of constants are 0. Moreover, y_{15} must be negative at point P. The value v of the objective function at P is $c_8 + d$. Since $c_8 > 0$, $v > d$.

Now, let us compute v from the first tableau. Since each entry in the objective row that comes before the entry under y_{15} must be non-positive (because the pivot took place in the column under y_{15}), the contribution of the term involving that entry is non-positive. The entry under y_{15} is positive, but, as we noted above, y_{15} has a negative value, and, therefore the contribution of the corresponding term is negative. The remaining non-basic variables at the top of this tableau all come after y_{15} and so must have been non-basic when P was defined. So, they were assigned the value 0, and their terms contribute 0 to the value. Hence, the value v is strictly less than d. This contradicts our earlier conclusion that $v > d$. This contradiction shows that Bland's Rules work.

3.7 Duality and the simplex method

A standard maximization lpp has the form:

$$\text{Maximize } u = c_1 x_1 + \cdots + c_n x_n + d$$

subject to constraints

$$\sum_{j=1}^{n} a_{ij} x_j \leq d_i \quad \text{for } 1 \leq i \leq m$$

Using slack variables r_i, we can replace the constraints by a standard canonical lpp with the equations $\sum_{j=1}^{n} a_{ij}x_j - d_i = -r_i$ for $1 \leq i \leq m$. The resulting simplex tableau is:

x_1	x_2	...	x_n	1	
a_{11}	a_{12}	...	a_{1n}	$-d_1$	$= -r_1$
a_{21}	a_{22}	...	a_{2n}	$-d_2$	$= -r_2$
...	...	...	...	...	....
a_{m1}	a_{m2}	...	a_{mn}	$-d_m$	$= -r_m$
c_1	c_2	...	c_n	d	$= u$

This maximization problem has a dual standard minimization lpp:

$$\text{Minimize } w = d_1y_1 + \cdots + d_my_m + d$$

subject to constraints

$$\sum_{i=1}^{m} a_{ij}y_i \geq c_j \qquad \text{for } 1 \leq j \leq n$$

Inserting slack variables s_j, we can replace the constraints by a standard canonical lpp with the equations $\sum_{i=1}^{m} a_{ij}y_i - c_j = s_j$ for $1 \leq j \leq n$. To obtain a tabular representation of this dual lpp, we need only extend the simplex tableau by adding suitable borders to the left and below:

	x_1	x_2	...	x_n	1	
y_1	a_{11}	a_{12}	...	a_{1n}	$-d_1$	$= -r_1$
y_2	a_{21}	a_{22}	...	a_{2n}	$-d_2$	$= -r_2$
...	...	...	...	...	...	....
y_m	a_{m1}	a_{m2}	...	a_{mn}	$-d_m$	$= -r_m$
-1	c_1	c_2	...	c_n	d	$= u$
	s_1	s_2	...	s_n	$-w$	

The expressions for the variables s_j of the new constraints and for w can be read off from this tableau by multiplying the left-most column of variables y_i (and -1) by the appropriate column.

The simplex method, when applied to the tableau for the maximization lpp, consists of a sequence of pivots. Each pivot exchanges the variables above and to the right of the pivot element. Such a pivot also can be thought of as determining an exchange of the two variables to the left of and below the pivot element. As a result, it gives rise to new expressions for the other variables in the lowest row in terms of the variables along the left side of the tableau. What is remarkable is that these expressions are exactly the equations that can be read off from the new tableau generated in the usual way by the pivot. We shall illustrate this in the following example.

Example 3.20

Consider the standard maximization lpp:

$$\text{Maximize } u = 2x_1 - 3x_2 + 5x_3 + 1$$

subject to

$$3x_1 + x_2 - x_3 \le 2 \quad \text{and} \quad 2x_1 - x_2 + 2x_3 \le 1$$

The dual lpp is:

$$\text{Minimize } w = 2y_1 + y_2 + 1$$

subject to

$$3y_1 + 2y_2 \ge 2, \; y_1 - y_2 \ge -3, \quad \text{and} \quad -y_1 + 2y_2 \ge 5$$

The extended tableau is:

	x_1	x_2	x_3	1	
y_1	3	1	-1	-2	$= -r_1$
y_2	2	-1	2	-1	$= -r_2$
-1	2	-3	5	1	$= u$
	s_1	s_2	s_3	$-w$	

By Bland's Rules, the first pivot is with respect to $a_{21} = 2$ and yields the tableau below. Note that we also have exchanged y_2 and s_1.

	r_2	x_2	x_3	1	
y_1	-3/2	5/2	-4	-1/2	$= -r_1$
s_1	1/2	-1/2	1	-1/2	$= -x_1$
-1	-1	-2	3	2	$= u$
	y_2	s_2	s_3	$-w$	

Originally we had $s_1 = 3y_1 + 2y_2 - 2$, and solving for y_2 yields $y_2 = -(3/2)y_1 + (1/2)s_1 + 1$, which, alternatively, can be read off from the new tableau by multiplying the left column by the column above y_2.

Similarly, $s_2 = y_1 - y_2 + 3 = y_1 - ((-3/2)y_1 + (1/2)s_1 + 1) + 3 = (5/2)y_1 - (1/2)s_1 + 2$, which also could be read off from the new tableau by multiplying the left column by the column above s_2.

We leave it as a tedious exercise to verify that what we have just noticed also holds true in the general case. The simplex method starts with a perfect canonical lpp. If the lpp has a solution, then the simplex method ends with a tableau that exhibits the solution and the maximum value of the objective function. If we use the extended tableaux, let us see what the result tells us about the dual lpp. Let us look at a concrete case so as to make the situation clear. The final tableaux might appear as follows (with inessential entries omitted):

	x_2	x_4	r_1	r_4	1	
y_1					−4	$=-x_1$
y_2					−1	$=-x_3$
y_4					−1	$=-r_2$
s_2					−3	$=-r_3$
−1	−5	−3	−3	−2	7	$=u$
	y_3	s_1	s_3	s_4	$-w$	

The maximum value of the objective function u is 7, attained when $x_2 = x_4 = r_1 = r_4 = 0$, and, therefore, $x_1 = 4$, and $x_3 = 1$. The objective function for the minimization lpp is w and, from the table, we see that $-w = -4y_1 - y_2 - y_4 - 3s_2 - 7$, and, therefore, $w = 4y_1 + y_2 + y_4 + 3s_2 + 7$. So, the minimum value of w occurs when $y_1 = y_2 = y_4 = s_2 = 0$ and $y_3 = 5$.

From this example we can see that the following result holds in general.

Fundamental theorem of duality
When the maximization lpp has a solution, so does its dual and the maximum value of the former is the minimum value of the latter. Moreover, when the dual has a solution, so does the original lpp.

Example 3.20   (Continued)
The second pivot element is 1. Straightforward computation yields the tableau:

	r_2	x_2	x_1	1	
y_1	1/2	1/2	4	−5/2	$=-r_1$
s_3	1/2	−1/2	1	−1/2	$=-x_3$
−1	−5/2	−1/2	−3	7/2	$=u$
	y_2	s_2	s_1	$-w$	

This concludes the pivoting. The common maximum of u and minimum of w is 7/2, attained at $x_1 = x_2 = 0$, $x_3 = {}^1\!/_2$, $y_1 = 0$, $y_2 = 5/2$.

Exercise 3.10
Write the dual of the given standard maximization lpp and find the common optimal values (if any) and where they are attained.

1. Maximize $u = 3x_1 - 4x_2 + 3$, subject to $2x_1 + 5x_2 \le 4$ and $x_1 - 3x_2 \le 1$.
2. Maximize $u = 2x_1 + x_2 - 4x_3 + 1$, subject to $x_1 - x_2 - x_3 \le 3$ and $2x_1 + 2x_2 + x_3 \le 4$.

3.8   Solution of game matrices

Using our earlier procedure for reducing von Neumann's Theorem to the Duality Theorem, we can solve matrix games by means of the simplex method. Given an $m \times n$ game matrix (b_{ij}), we add, if necessary, a constant K to all entries so as to obtain a matrix (a_{ij}) with all entries positive. Our aim is to solve the lpp:

$$\text{Maximize } \sum_{j=1}^{n} \beta_j$$

subject to

$$\sum_{j=1}^{n} a_{ij}\beta_j \leq 1 \qquad \text{for } i = 1, \ldots, m$$

and its dual lpp:

$$\text{Minimize } \sum_{i=1}^{m} \alpha_i$$

subject to

$$\sum_{i=1}^{m} a_{ij}\alpha_i \geq 1 \qquad \text{for } j = 1, \ldots, n$$

If these lpp's are solvable, their common optimal value v^* will be positive, and an equilibrium pair (X, Y) is given by $X = (x_1, \ldots, x_m)$, $Y = (y_1, \ldots, y_n)$, with $x_i = v^*\alpha_i$ and $y_j = v^*\beta_j$. The value v of the game would be $v^* - K$.

Example 3.21
Consider the 3×3 game matrix

$$\begin{pmatrix} 3 & -1 & 0 \\ 2 & 4 & 1 \\ 1 & 2 & 3 \end{pmatrix}$$

Add 2 to all entries so that all entries will become positive.

$$\begin{pmatrix} 5 & 1 & 2 \\ 4 & 6 & 3 \\ 3 & 4 & 5 \end{pmatrix}$$

So we must solve the lpp's:

$$\text{Maximize } \beta_1 + \beta_2 + \beta_3$$

subject to the constraints

$$5\beta_1 + \beta_2 + 2\beta_3 \leq 1$$

$$4\beta_1 + 6\beta_2 + 3\beta_3 \leq 1, \quad \text{and}$$

$$3\beta_1 + 4\beta_2 + 5\beta_3 \leq 1$$

and its dual

$$\text{Minimize } \alpha_1 + \alpha_2 + \alpha_3$$

subject to the constraints

$$5\alpha_1 + 4\alpha_2 + 3\alpha_3 \geq 1$$

$$\alpha_1 + 6\alpha_2 + 4\alpha_3 \geq 1, \quad \text{and}$$

$$2\alpha_1 + 3\alpha_2 + 5\alpha_3 \geq 1$$

Introducing slack variables r_1, r_2, r_3 for the maximization problem, we obtain the constraint equations

$$5\beta_1 + \beta_2 + 2\beta_3 + r_1 = 1$$

$$4\beta_1 + 6\beta_2 + 3\beta_3 + r_2 = 1, \quad \text{and}$$

$$3\beta_1 + 4\beta_2 + 5\beta_3 + r_3 = 1$$

On the other hand, we can introduce slack variables s_1, s_2, s_3 for the minimization problem such that

$$5\alpha_1 + 4\alpha_2 + 3\alpha_3 - s_1 = 1$$

$$\alpha_1 + 6\alpha_2 + 4\alpha_3 - s_2 = 1, \quad \text{and}$$

$$2\alpha_1 + 3\alpha_2 + 5\alpha_3 - s_3 = 1$$

Then we get the extended simplex tableau

	β_1	β_2	β_3	1	
α_1	5	1	2	−1	$= -r_1$
α_2	4	6	3	−1	$= -r_2$
α_3	3	4	5	−1	$= -r_3$
−1	1	1	1	0	$= u$
	s_1	s_2	s_3	−w	

Bland's Rules tell us to pivot with respect to $a_{11} = 5$. (We take the ordering of variables to be $\beta_1, \beta_2, \beta_3, r_1, r_2, r_3$.) This yields the new tableau:

	r_1	β_2	β_3	1	
s_1	1/5	1/5	2/5	−1/5	$= -\beta_1$
α_2	−4/5	26/5	7/5	−1/5	$= -r_2$
α_3	−3/5	17/5	19/5	−2/5	$= -r_3$
−1	−1/5	4/5	3/5	1/5	$= u$
	α_1	s_2	s_3	−w	

The next pivot is with respect to 26/5, yielding:

	r_1	r_2	β_3	1	
s_1	3/13	−1/26	9/26	−5/26	$= -\beta_1$
s_2	−2/13	5/26	7/26	−1/26	$= -\beta_2$
α_3	−1/13	−17/26	75/26	−7/26	$= -r_3$
−1	−1/13	−2/13	5/13	3/13	$= u$
	α_1	α_2	s_3	−w	

Our next pivot is with respect to 75/26 and yields:

	r_1	r_2	r_3	1	
s_1	6/25	38/325	−3/25	−4/25	$= -\beta_1$
s_2	−11/75	38/975	−7/75	−1/75	$= -\beta_2$
s_3	−2/75	−17/75	26/75	−7/75	$= -\beta_3$
−1	−1/15	−1/15	−2/15	4/15	$= u$
	α_1	α_2	α_3	−w	

We have found a solution: $\beta_1 = 4/25$, $\beta_2 = 1/75$, $\beta_3 = 7/75$, $\alpha_1 = 1/15$, $\alpha_2 = 1/15$, $\alpha_3 = 2/15$. The common optimal value is 4/15, and its reciprocal 15/4 is the value v^* of the game. The equilibrium pair (X, Y) is given by $X = (x_1, x_2, x_3)$ and $Y = (y_1, y_2, y_3)$, where $x_i = v^* \alpha_i$ and $y_j = v^* \beta_j$. Thus, $X = (1/4, 1/4, 1/2)$ and $Y = (3/5, 1/20, 7/20)$.

Since we added 2 to all entries of the original game, the value of that game is $v = (15/4) - 2 = 7/4$. It is easy to verify that $u(X) = 7/4 = w(Y)$, which confirms that (X, Y) is an equilibrium pair.

Example 3.22
Let us solve the 3×4 game matrix

$$\begin{pmatrix} 2 & -3 & 1 & 5 \\ 1 & 4 & 3 & -1 \\ 2 & 0 & -2 & 4 \end{pmatrix}$$

Add 4 to all the entries.

$$\begin{pmatrix} 6 & 1 & 5 & 9 \\ 5 & 8 & 7 & 3 \\ 6 & 4 & 2 & 8 \end{pmatrix}$$

With suitable slack variables, we obtain the simplex tableau

	β_1	β_2	β_3	β_4	1	
α_1	6	1	5	9	−1	$=-r_1$
α_2	5	8	7	3	−1	$=-r_2$
α_3	6	4	2	8	−1	$=-r_3$
−1	1	1	1	1	0	$=u$
	s_1	s_2	s_3	s_4	$-w$	

The first pivot is with respect to $a_{11} = 6$. (The order of the variables is assumed to be $\beta_1, \beta_2, \beta_3, \beta_4, r_1, r_2, r_3$.) So we get the tableau:

	r_1	β_2	β_3	β_4	1	
s_1	1/6	1/6	5/6	3/2	−1/6	$=-\beta_1$
α_2	−5/6	43/6	17/6	−9/2	−1/6	$=-r_2$
α_3	−1	3	−3	−1	0	$=-r_3$
−1	−1/6	5/6	1/6	−1/2	1/6	$=u$
	α_1	s_2	s_3	s_4	$-w$	

The next pivot element is $a_{32} = 3$. The resulting tableau is:

	r_1	r_3	β_3	β_4	1	
s_1	2/9	−1/18	1	14/9	−1/6	$=-\beta_1$
α_2	14/9	−43/18	10	−19/9	−1/6	$=-r_2$
s_2	−1/3	1/3	−1	−1/3	0	$=-\beta_2$
−1	1/9	−5/18	1	−2/9	1/6	$=u$
	α_1	α_3	s_3	s_4	$-w$	

Now we pivot with respect to $a_{23} = 10$.

	r_1	r_3	r_2	β_4	1	
s_1	1/15	11/60	−1/10	53/30	−3/20	$=-\beta_1$
s_3	7/45	−43/180	1/10	−19/90	−1/60	$=-\beta_3$
s_2	−8/45	17/180	1/10	−49/90	−1/60	$=-\beta_2$
−1	−2/45	−7/180	−1/10	−1/90	11/60	$=u$
	α_1	α_3	α_2	s_4	$-w$	

This yields a solution: $\alpha_1 = 2/45$, $\alpha_2 = 1/10$, $\alpha_3 = 7/180$, $\beta_1 = 3/20$, $\beta_2 = 1/60$, $\beta_3 = 1/60$, $\beta_4 = 0$. The common optimal value is 11/60. Hence, its reciprocal 60/11 is the value v^* of the game matrix. The equilibrium pair is (X, Y), where $X = (x_1, x_2, x_3)$ and $Y = (y_1, y_2, y_3, y_4)$, with $x_i = v^*\alpha_i$ and $y_j = v^*\beta_j$. Thus, $X = (8/33, 6/11, 7/33)$ and $Y = (9/11, 1/11, 1/11, 0)$. The value v of the original game matrix is $v^* - 4 = 16/11$. It is easy to verify that $u(X) = 16/11 = w(Y)$, which confirms that (X, Y) is an equilibrium pair.

Exercise 3.11
Solve the following game matrices.

$$(a) \begin{pmatrix} 0 & -1 & 1 & 1 \\ 2 & 0 & 0 & 1 \\ 0 & 1 & 0 & -1 \end{pmatrix} \quad (b) \begin{pmatrix} 1 & -2 & -3 & 2 \\ 2 & 3 & 4 & 0 \end{pmatrix}$$

Exercise 3.12
Colonel Blotto Games. These are two-person games between Colonels Alexander (A) and Blotto (B). Each colonel commands a certain number of regiments and they both want to take over two forts (I) and (II). Each colonel sends some of his regiments to occupy Fort (I) and the rest to occupy Fort (II). Points are awarded to the colonels according to the following rules. If more regiments are sent to a particular fort by one colonel than by the other, then that colonel receives one point for control of the fort plus one point for each of the opponent's regiments captured at that fort. If the colonels send the same number of regiments to a given fort, neither receives any points for that fort. A strategy for a colonel consists of a pair (j, k), where j and k are the numbers of regiments sent by the colonel to forts (I) and (II), respectively. Each colonel wants to maximize his total number of points.

(a) Assume that (A) has 4 regiments and (B) has 3. Verify that the following is the game matrix, where, as usual, each entry is the gain for player (A), and then find an equilibrium pair (X, Y) for the game, and the value v of the game.

	(0,3)	(1,2)	(2,1)	(3,0)
(0,4)	4	2	1	0
(1,3)	1	3	0	-1
(2,2)	-2	2	2	-2
(3,1)	-1	0	3	1
(4,0)	0	1	2	4

(b) The same as Part (a) except that (A) has 3 regiments and (B) has 2 regiments, and you have to construct the game matrix yourself.

Proofs of Facts 1–4.

x_1	.....	x_n	1	
a_{11}	.....	a_{1n}	$-b_1$	$= -x_{n+1}$
a_{21}	.....	a_{2n}	$-b_2$	$= -x_{n+2}$
...	.....	...	...	......
a_{m1}	.....	a_{mn}	$-b_m$	$= -x_{n+m}$
c_1	.....	c_n	d	$= u$

Fact 1.   If all the entries a_{ij} in the column above a positive entry c_j are negative or zero, then the lpp has no solution.

Proof. $u = c_1 x_1 + \cdots + c_j x_j + \cdots + c_n x_n + d$. Assign the value 0 to all variables except x_j and assign a positive value K to x_j. Then $u = c_j K + d$. Since $c_j > 0$, u can be made arbitrarily large by choosing K sufficiently large, and u would have no maximum. It only remains to check that the points yielding the arbitrarily large values are feasible.

Since $x_1, \ldots, x_n$ are all non-negative, we must show that $x_{n+1}, \ldots, x_{n+m}$ also are non-negative. Look, for example, at x_{n+1}. At the assigned point, the first row of the tableau gives $-x_{n+1} = a_{1j} K - b_1$. Since we are dealing with a perfect canonical lpp, b_1 is non-negative, and, by hypothesis, $a_{1j} \leq 0$. Hence, $a_{1j} K - b_1 \leq 0$, and, therefore, $x_{n+1} \geq 0$. The same argument holds for all of the variables $x_{n+1}, \ldots, x_{n+m}$, and so the points we are using are all feasible.

Fact 2.   When we carry out

Step 2.   Pivot with respect to that positive a_{ij} for which the ratio b_i / a_{ij} is minimal,

1. The result is a perfect standard canonical lpp, that is, the numbers in the column of constants in the resulting tableau are negative or zero.
2. The value of the objective function at the new basic point is greater than or equal to its value at the previous basic point. Let the ith row be the pivot row. If b_i is positive, the value of the objective function is strictly bigger than before. When $b_i = 0$ (this is called the *degenerate case*), then the value of the objective function does not change. Moreover, in the degenerate case, the entire column of constants remains the same, and the new basic point is the same as the previous basic point.

Proof

1. Consider any entry Z in the column of constants of the new tableau, say, in the kth row. We must show that $Z \leq 0$. The given tableau is as follows, where the pivot element is a_{ij}.

x_1.....	x_j	..... x_n	1	
.....	..	.....	..	....
.....	a_{kj}	.....	$-b_k$	$= -r_k$
.....	..	.....	..	....
.....	a_{ij}	.....	$-b_i$	$= -r_i$
.....	..	.....	..	....
c_1.....	c_j	..... c_n	d	$= u$

After the pivot, the new tableau looks like:

x_1.....	r_j	..... x_n	1	
.....	..	.....	..	....
.....		.....	Z	$= -r_k$
.....	..	.....	..	....
.....	$1/a_{ij}$	.....	$-b_i/a_{ij}$	$= -x_j$
.....	..	.....	..	....
c_1.....	c_j	..... c_n	d	$= u$

(i) Since $a_{ij} > 0$ and $b_i \geq 0$, the constant in the ith row, $-b_i/a_{ij}$, is negative or zero.

(ii) Note that, after the pivot, $-x_j = (a_{i1}/a_{ij})x_1 + \cdots + (1/a_{ij})r_i + \cdots + (a_{in}/a_{ij})x_n - (b_i/a_{ij})$. The new kth row gives us: $a_{k1}x_1 + \cdots + (-a_{kj}(-x_j)) + \cdots + a_{kn}x_n + (-b_k) = -r_k$. Substituting for $-x_j$, we see that the constant Z in the new kth row is $(a_{kj}b_i/a_{ij}) - b_k$.

Case 1. $a_{kj} \leq 0$. Then $(a_{kj}b_i)/a_{ij} \leq 0$. So, $Z \leq 0$.

Case 2. $a_{kj} > 0$. By the pivoting rules for the minimality of b_i/a_{ij}, we have $b_i/a_{ij} \leq b_k/a_{kj}$. Then $(a_{kj}b_i)/a_{ij} \leq b_k$ and, therefore, $Z = (a_{kj}b_i/a_{ij}) - b_k \leq 0$.

2.   After the pivot, we evaluate the objective function at the new basic point when $r_i = 0$ and $x_k = 0$ for $k = 1, \ldots, j-1, j+1, \ldots, n$. Since $u = c_1 x_1 + \cdots + (-c_j)(-x_j) + \cdots + x_n + d = c_1 x_1 + \cdots + (-c_j)((a_{i1}/a_{ij})x_1 + \cdots + (1/a_{ij})r_i + \cdots + (a_{in}/a_{ij})x_n - (b_i/a_{ij})) + \cdots + c_n x_n + d$, the value of u at the new point is $(c_j b_i/a_{ij}) + d$. When $b_i > 0$, this new value is greater than the value d at the earlier basic point where $x_k = 0$ for $1 \leq k \leq n$. When $b_i = 0$, the new value is d again. Let us also show that, in this degenerate case when $b_i = 0$, all the entries in the column of constants of the new tableau are the same as they were before. In fact, in an arbitrary kth row, we get $-r_k = a_{k1}x_1 + \cdots + a_{k,j-1}x_{j-1} + (-a_{kj})((a_{i1}/a_{ij})x_1 + \cdots + (1/a_{ij})r_i + \cdots + (a_{in}/a_{ij})x_n - (b_i/a_{ij})) + \cdots + a_{kn}x_n - b_k$, so that the entry in the kth row in the column of constants is $(a_{kj}b_i/a_{ij}) - b_k = -b_k$, which was the constant in the kth row in the previous tableau.

Fact 3.   There is a way of modifying the simplex method so that it never goes on forever. We already have seen that Bland's Rules provide such a modification.

Fact 4.   In the Big M method, if there is a solution, then the artificial variables all have the value zero and we obtain a solution of the original lpp.

Proof.   Let $u(x_1, \ldots, x_n)$ be the original objective function and let the constraints, after slack variables are introduced, have the form $g_i(x_1, \ldots, x_n) + r_i = d_i$. When $d_i < 0$, the Big M method introduces a new artificial variable.

Let $A_1, \ldots, A_k$ be the artificial variables introduced via the Big M method. Then the new Big M constraints have the form $-g_i(x_1, \ldots, x_n) - r_i + A_i = -d_i$. We assume that there is a feasible point $(a_1, \ldots, a_n)$ for the original lpp. In the course of finding the Big M solution, the only limitations on M occurred when, in the objective row, entries that were linear functions of M had to be negative. (For example, if an entry was $20 - 3M$, then M had to be greater than 20/3.)

Since this happened only a finite number of times, we could determine a number such that any value of M greater than that number would suffice. The Big M method yields the same solution $(a_1^*, \ldots, a_n^*, A_1^*, \ldots, A_k^*)$ for the variables when M had any such value. That Big M solution yields the maximum value for the Big M objective function $u(x_1, \ldots, x_n) - MA_1 - \cdots - MA_k$. But the point $(a_1, \ldots, a_n, 0, \ldots, 0)$, where all A_i are set equal to 0, satisfies the Big M constraints. The value of the objective function $u(x_1, \ldots, x_n) - MA_1 - \cdots - MA_k$ at that point is $u(a_1, \ldots, a_n)$. Since the maximum of that objective function is reached at $(a_1^*, \ldots, a_n^*, A_1^*, \ldots, A_k^*)$, we obtain $u(a_1, \ldots, a_n) \leq u(a_1^*, \ldots, a_n^*) - MA_1^* - \cdots - MA_k^*$. Now, if any A_i^* is positive, then by making M large enough, the right-hand side of this inequality can be made smaller than the left-hand side $u(a_1, \ldots, a_n)$. Therefore, every A_i^* is zero.

chapter four

Non-zero-sum games and k-person games

4.1 The general setting

The theory of two-person zero-sum games that was studied in Chapter 2 was relatively clear, precise, and successful. When we enlarge our study to k-person games, for $k > 2$, or when we lift the restriction to zero-sum games, the conceptual playing field becomes much harder to survey and conclusive results are sometimes harder or impossible to obtain. Unfortunately, most of the important real-life games are either k-person games with $k > 2$ or non-zero-sum. Nevertheless, not all is lost. Analogues of some of the notions and results of the two-person zero-sum theory can be found and will be developed in this chapter. Moreover, the variety and intricacy of the problems encountered in the larger setting have offered and still offer a host of interesting challenges.

As in Chapter 2, we shall deal with games in normal form. We shall assume that each player in a game has a finite number of strategies. If we designate the players by $\mathbf{p}_1,\ldots, \mathbf{p}_k$, and if $S_1,\ldots, S_k$ are any strategies for players $\mathbf{p}_1,\ldots, \mathbf{p}_k$, respectively, then we shall denote by $P_j(S_1,\ldots, S_k)$ the (expected) pay-off for player $\mathbf{p}_j$ when $\mathbf{p}_1,\ldots, \mathbf{p}_k$ play strategies $S_1,\ldots, S_k$, respectively.

If there are just two players, say A and B, then we can represent a game by a matrix of pay-offs, where each entry in the matrix is a pair (a,b) such that a is the pay-off for A and b is the pay-off for B.

Example 4.1

If A has three strategies A_1, A_2, A_3 and B has two strategies B_1 and B_2, then the following matrix defines a game between A and B.

	B_1	B_2
A_1	$(-1,5)$	$(0,2)$
A_2	$(3,1)$	$(1,0)$
A_3	$(1/2,2)$	$(4,1)$

For example, the entry (3,1) indicates that, if A plays strategy A_2 and B plays strategy B_1, then A has a pay-off of 3 and B has a pay-off of 1. Note that this game is not zero-sum. For instance, the sum for (3,1) is 4 and the sum for (1,0) is 1.

It is sometimes useful to break up a game matrix into two separate matrices, one showing the pay-offs for player A and the other showing the pay-offs for player B. In Example 4.1, we would obtain:

	B_1	B_2			B_1	B_2
A_1	−1	0		A_1	5	2
A_2	3	1		A_2	1	0
A_3	1/2	4		A_3	2	1

(Pay-offs for A)   (Pay-offs for B)

In this case, we can try to use reasoning similar to that used in Chapter 2. Looking at the matrix for B's pay-offs, we see that strategy B_1 dominates strategy B_2, that is, playing B_1 always yields a larger pay-off for B than playing strategy B_2. Hence, B will play B_1. Now, looking at the matrix for A's pay-offs, A should consider only the pay-offs in the column under B_1, since A can assume that B will play B_1. The highest pay-off for A would then be 3, when A plays A_2. Thus, A should play A_2 and B should play B_1. Then A will receive 3 and B will receive 1.

As you would expect, not all games with two players are susceptible to such a simple analysis as in Example 4.1. We will have to study what happens in the general case.

When there are three players in a game, a pay-off matrix representing the game would require three dimensions and would be difficult to draw. When there are more than three players, no tabular picture is possible at all. Therefore, although our theoretical discussions will deal with the general case of any finite number of players, most of our examples will involve only two players.

Let us consider any two-person game, with players A and B. Let A's strategies be $A_1, \ldots, A_m$, and let B's strategies be $B_1, \ldots, B_n$. The game matrix will be an $m \times n$ matrix whose entries are pairs (c_{ij}, d_{ij}) such that c_{ij} is the pay-off $P_1(A_i, B_j)$ for A when A plays strategy A_i and B plays strategy B_j and d_{ij} is the pay-off $P_2(A_i, B_j)$ for B when A plays strategy A_i and B plays strategy B_j.

Player A has at least one maximin strategy, defined as in Chapter 2. Recall that, for each row of the pay-off matrix, we find the minimum that A can receive in that row. Then the *maximin for A* is the maximum of all such minima, taken over all rows.

A strategy for any row that yields this maximin for player A is called a *maximin strategy for A*. Remember that A can be guaranteed a pay-off of at least his maximin by playing a maximin strategy for A. Likewise, player B has at least one maximin strategy. For each column of the pay-off matrix, we find the minimum that B can receive in that column. Then the maximin for B is the maximum of all such minima, taken over all columns.

A strategy for any column that yields this maximin for B is called a *maximin strategy for B*. B is guaranteed a pay-off of at least his maximin by playing a maximin strategy for B. If A_i is a maximin strategy for A and B_j is a maximin strategy for B, then (A_i, B_j) will be called a *maximin pair*.[1] The pay-offs that A and B get when they play a maximin pair are at least as big as their maximins.

In Example 4.1, the maximin for A is 1 and the only maximin strategy for A is A_2. The maximin for B is 1 and the only maximin strategy for B is B_1. Thus, the only maximin pair is (A_2, B_1). For this maximin pair, A gets the pay-off 3 and B gets the pay-off 1. Thus, A gets more than his maximin, whereas B gets exactly his maximin.

Example 4.2
Consider the game determined by the following matrix.

	B_1	B_2
A_1	(1,3)	(3,5)
A_2	(2,1)	(0,2)

The maximin for A is 1 and the only maximin strategy is A_1. The maximin for B is 2 and the only maximin strategy for B is B_2. Hence, the maximin pair is (A_1, B_2), with pay-offs of 3 for A and 5 for B. So, both players get more than their maximins. (Note that, in this example, the players can be led to the maximin pair by dominance arguments.)

Example 4.3
Let us look at the game defined by the matrix

	B_1	B_2
A_1	(1,2)	(3,6)
A_2	(4,4)	(2,3)

The maximin for A is 2 and A's only maximin strategy is A_2. The maximin for B is 3 and B's only maximin strategy is B_2. Hence, the maximin pair is (A_2, B_2) with pay-offs 2 and 3 for A and B, respectively, and A and B would receive exactly their maximins.

Example 4.4
Consider the game matrix

	B_1	B_2	B_3
A_1	(1,2)	(4,3)	(2,1)
A_2	(2,4)	(−1,2)	(3,5)
A_3	(2,2)	(1,4)	(2,0)

[1] Note that minimax values are not important here, since the game may not be zero-sum.

A's maximin is 1, but there are two maximin strategies for A, A_1 and A_3. B's maximin is 2, with maximin strategies B_1 and B_2. So, there are four maximin pairs, (A_1,B_1), (A_3,B_1), (A_1,B_2), (A_3,B_2), with pay-offs (1,2), (2,2), (4,3), and (1,4), respectively. Assuming that A will not play strategy A_2, player B would prefer strategy B_2 over B_1 on the basis of dominance. Given that B plays B_2, player A would choose strategy A_1. Note that the resulting pair of strategies (A_1,B_2) is *stable*, in the sense that neither player would benefit by changing that player's strategy. *Observe that the pay-offs (4,3) determined by (A_1,B_2) has a first component 4 that is the maximum in its column, and a second component 3 that is the maximum in its row. Clearly, having a first component that is the maximum in its column and a second component that is the maximum in its row is a necessary and sufficient condition for a pair to be stable.* The other three maximin pairs are not stable. (A_2,B_3) is also a stable pair, but there is no apparent reasoning that would lead to it. Note that the original matrix had no dominance relations for A or B.

Exercise 4.1
In Examples 4.1 to 4.3, are the maximin pairs stable?

Exercise 4.2
For the following game matrices, find the maximins and the maximin strategies for A and B, and find the maximin pairs. Determine which pairs of strategies in the original matrix are stable.

(a)

	B_1	B_2	B_3
A_1	(−1,0)	(3,2)	(1,4)
A_2	(2,5)	(1,4)	(2,3)

(b)

	B_1	B_2
A_1	(1,2)	(0,3)
A_2	(−1,3)	(1,1)

(c)

	B_1	B_2	B_3
A_1	(5,2)	(3,4)	(−2,3)
A_2	(3,2)	(1,2)	(4,2)
A_3	(0,4)	(2,2)	(1,5)

The notions of *maximin* and *maximin strategy* can be generalized to games with more than two players. Let A be a player. Consider any strategy

A_i for A and find the minimum pay-off for A when A plays strategy A_i against all possible choices of strategies for the other players. Then the *maximin for A* is the maximum of such minima over all strategies A_i for A. A *maximin strategy for A* is a strategy for A that yields the maximin for A. If there are k players, then a maximin k-tuple is defined to be a k-tuple such that, for $1 \leq i \leq k$, its ith component is a maximin strategy for the ith player. When $k = 2$, a maximin k-tuple is just what we previously called a maximin pair.

4.2 Nash equilibria

The key notion that we will need is a natural generalization of the concepts of saddle point and equilibrium pair in Chapter 2. Let us assume that a game with k players $p_1, \ldots, p_k$ is given in normal form. By a *Nash equilibrium* we shall mean a k-tuple of strategies $(S_1, \ldots, S_k)$, where each S_i is a strategy for p_i, and such that, if all but one of the players retain their strategies in this k-tuple, then the remaining player cannot improve his pay-off by changing to another one of his strategies.[2]

When $k = 2$, a Nash equilibrium is just a pair of strategies that is stable in the sense used above in Example 4.4. Although Nash equilibria seem to involve just a natural generalization of the notion of equilbrium pair in two-person zero-sum games, we shall see that they lack some of the latter's nice properties.

Before studying Nash equiilibria more closely, let us look at a few competing notions. Let us assume as before that a game with k players $p_1, \ldots, p_k$ is given in normal form. If $(S_1, \ldots, S_k)$ is a k-tuple of strategies, where each S_i is a strategy for player p_i, then there is a corresponding k-tuple $(c_1, \ldots, c_k)$ of numbers such that each c_i is the pay-off for player p_i when the players play the respective strategies $S_1, \ldots, S_k$. Given two such k-tuples of pay-offs $(a_1, \ldots, a_k)$ and $(b_1, \ldots, b_k)$, we say that $(a_1, \ldots, a_k)$ is *better than* $(b_1, \ldots, b_k)$ if the following two conditions hold:

1. $a_i \geq b_i$ for all i
2. $a_j > b_j$ for at least one j (or, equivalently, $(a_1, \ldots, a_k)$ is not the same as $(b_1, \ldots, b_k)$).

Definitions

A k-tuple of strategies $(S_1, \ldots, S_k)$ is said to be *Pareto-optimal*[3] if no k-tuple of pay-offs determined by any other k-tuple of strategies is better than the k-tuple of pay-offs determined by $(S_1, \ldots, S_k)$. A k-tuple of strategies $(S_1, \ldots, S_k)$ is said to be *best* if the k-tuple of pay-offs determined by $(S_1, \ldots, S_k)$ is better than the k-tuple of pay-offs determined by any other k-tuple of strategies. A k-tuple of strategies $(S_1, \ldots, S_k)$ is said to be *among-the-best* if the k-tuple

---

[2]In economics, what is known as a *Cournot equilibrium* is essentially a Nash equilibrium.
[3]This concept is named after the Italian sociologist Vilfredo Pareto (1848–1923).

of pay-offs determined by $(S_1, \ldots, S_k)$ is better than or identical to the k-tuple of pay-offs determined by any other k-tuple of strategies. (Clearly, if a k-tuple is best, then it is among-the-best.)

Nash equilibria, Pareto-optimal k-tuples, best, and among-the-best k-tuples are in one way or another desirable by the players. We shall examine some of their properties and relationships in what follows. In general, the concept of Nash equilibrium will have the most significance.

Example 4.5

Consider the two-person game determined by the matrix

$$
\begin{array}{ccc}
 & B_1 & B_2 \\
A_1 & (1,3) & (0,2) \\
A_2 & (2,1) & (-1,2)
\end{array}
$$

The pairs (A_1, B_1) and (A_2, B_1) are Pareto-optimal, but there is no among-the-best pair and there is no Nash equilibrium. The maximin for A is 0 and A_1 is the only maximin strategy for A. The maximin for B is 2, and B has maximin strategy B_2. So, (A_1, B_2) is a maximin pair.

Example 4.5 shows that a maximin k-tuple need not be a Nash equilibrium, Pareto-optimal, or among-the-best. It also shows that a Pareto-optimal k-tuple need not be among-the-best or a Nash equilibrium or maximin. Moreover, a game need not have any Nash equilibrium or an among-the-best k-tuple. However, the following result shows that this is not true for Pareto-optimality.

Theorem 4.1

Any game with k players has at least one Pareto-optimal k-tuple.

Proof.   Note first that, if a k-tuple K is better than a k-tuple L and L is better than a k-tuple M, then K is better than M. Assume now, for the sake of contradiction, that some game with k players has no Pareto-optimal k-tuple. Choose any k-tuple of strategies K_1. Since K_1 is not Pareto-optimal, there is a k-tuple K_2 that is better than K_1. Since K_2 is not Pareto-optimal, there is a k-tuple K_3 that is better than K_2 (and, therefore, also better than K_1). Continuing in this way, we see that there must be an infinite sequence of distinct k-tuples. But this is impossible because only finitely many k-tuples of strategies exist (since there are only k players and each player has only a finite number of strategies).   ∎

Exercise 4.3

Show that every among-the-best k-tuple also is a Nash equilibrium and Pareto-optimal.

Exercise 4.4
In Examples 4.1 to 4.4, find the Nash equilibria, the among-the-best pairs, and the Pareto-optimal pairs.

Exercise 4.5
Show by an example that a Nash equilibrium need not be Pareto-optimal.

Exercise 4.6
Find any Nash equilibria, Pareto-optimal pairs, and among-the-best pairs in the matrix

	B_1	B_2
A_1	(5,1)	(3,2)
A_2	(2,3)	(0,4)

There is a simple labeling technique for locating all Nash equilibria in two-person games. In each row, attach a sharp # to all second components that are maximal among all second components in that row. In each column, attach a sharp # to all first components that are maximal among all first components in that column. Then any pair that has a sharp attached to both components is a Nash equilibrium.

Exercise 4.7
Verify that this technique works in the following cases:

(a)

	B_1	B_2	B_3
A_1	(1,−1)	(2,0)	(0,3#)
A_2	(0,2)	(4#,1)	(1,3#)
A_3	(2#,1)	(0,0)	(5#,2#)
A_4	(−3,3#)	(1,2)	(0,1)

(b)

	B_1	B_2	B_3
A_1	(1#,2#)	(−1,0)	(0,2#)
A_2	(1#,1)	(2#,2)	(3,3#)
A_3	(0,3#)	(0,2)	(4#,3#)

(c)

	B_1	B_2
A_1	(5#,3#)	(−1,2)
A_2	(2,0)	(1#,1#)

(d)

	B_1	B_2	B_3
A_1	(0,1)	(5#,4)	(1,7#)
A_2	(1#,2)	(4,3#)	(2#,0)

Exercise 4.8
Prove that, in general, the pairs selected by this technique are always the same as the Nash equilibria.

Exercise 4.9
Use the labeling technique for locating Nash equilibria to find all such equilibria in Examples 4.1 to 4.5 and Exercises 4.2 and 4.6.

Recall that, for a two-person zero-sum game, any two saddle points yielded the same pay-off. However, this is not necessarily true for Nash equilibria of two-person non-zero-sum games. This can be observed above in Examples 4.3 and 4.4, and in Exercises 4.2(c) and 4.7(b,c). Similarly, in a two-person zero-sum game, if there are saddle points at two opposite corners of a rectangle, then the other two corners also must be saddle points. But the analogue does not hold for Nash equilibria in two-person non-zero-sum games, as can be seen in Examples 4.3 and 4.4, and in Exercises 4.2(c) and 4.7(b,c).

Remember that not every two-person zero-sum game has a saddle point (and, therefore, an equilibrium pair). That led to the introduction of mixed strategies and to the concept of equilibrium pairs of mixed strategies, and von Neumann's Theorem showed that every two-person zero-sum game had at least one equilibrium pair of mixed strategies. Now we already know that not every non-zero sum game has a Nash equilibrium. So, we shall imitate the tactics that were applied in zero-sum games.

Consider any k-person game with players $p_1,\ldots, p_k$. For any player p_i, let $A_1,\ldots, A_m$ be the strategies for that player. By a **mixed strategy** for player p_i we mean any m-tuple $(x_1,\ldots, x_m)$ of numbers such that $0 \le x_j \le 1$ for $1 \le j \le m$ and $x_1 + \cdots + x_m = 1$.

Assume now that $S_1,\ldots, S_k$ are mixed strategies for players $p_1,\ldots, p_k$, respectively. Then the expected pay-off $P_j(S_1,\ldots, S_k)$ for player p_j when $p_1,\ldots, p_k$ play mixed strategies $S_1,\ldots, S_k$, respectively, is defined to be the sum of all products of the form $x_{1r_1} x_{2r_2} \cdots x_{kr_k} P_j(A_{1r_1}, A_{2r_2},\ldots, A_{kr_k})$, where each A_{ir_i} is a strategy for player p_i, and x_{ir_i} is the corresponding coordinate in the strategy S_i.

To make the formula for the expected pay-off somewhat clearer, let us look at the case of a two-person game with players A and B. Let $A_1,\ldots, A_m$ be the strategies for player A, and let $B_1,\ldots, B_n$ be the strategies for player B. Let $X = (x_1,\ldots, x_m)$ be a mixed strategy for player A, and let $Y = (y_1,\ldots, y_n)$ be a mixed strategy for player B. Then the expected pay-off $P_A(X,Y)$ for player A is $\sum_{1 \le i \le m, 1 \le j \le n} x_i y_j\, P_A(A_i, B_j)$, and the expected pay-off $P_B(X,Y)$ for player B is $\sum_{1 \le i \le m, 1 \le j \le n} x_i y_j\, P_B(A_i, B_j)$. Note that the expected pay-offs $P_A(A_i, B_j)$ and $P_B(A_i, B_j)$ for players A and B, respectively, are already known for the pure strategies A_i and B_j.

As in the case of zero-sum games, the definition of equilibrium k-tuple of mixed strategies is an obvious extension of the definition for pure strategies. Assume that $S_1,\ldots, S_k$ are mixed strategies for players $p_1,\ldots, p_k$, respectively. Then $(S_1,\ldots, S_k)$ is said to be an *equilibrium k-tuple* if the following condition holds: For each player p_j, the expected pay-off for p_j with respect to $S_1,\ldots, S_k$ is at least as great as the expected pay-off for p_j with respect to the k-tuple that is the same as $S_1,\ldots, S_k$ except that S_j can be replaced by any other pure strategy for p_j. This means that no player can receive a better pay-off by changing his strategy in the k-tuple, while the other players play their same strategies in the k-tuple.

As in the case of two-person zero-sum games, it turns out that every k-person game has a mixed equilibrium k-tuple. This was proved by John F. Nash Jr. in 1949 (Nash, [1950]). The proof uses ideas that cannot be developed here; a sketch of the proof is given in Appendix 3.

Along with John C. Harsanyi and Reinhard Selten, Nash won the Nobel Prize in Economic Sciences in 1994 for work in game theory. In fact, Nash made even deeper contributions in other branches of mathematics. (See Milnor [1995].) However, his career was cut short by the onset of schizophrenia in the late 1950s. Eventually, beginning in the 1980s, his condition improved and he has experienced a moderately good recovery. All of this is described in the biography by Sylvia Nasar [1998], on which the 2001 film *A Beautiful Mind* is based.

Let us look at some examples where there is no Nash equilibrium consisting of pure strategies and where we must find ways to find a mixed Nash equilibrium.

Example 4.6
Consider the game with the matrix

		y B_1	$1-y$ B_2
x	A_1	(2,5)	(5,2)
$1-x$	A_2	(3,0)	(1,3)

Here, player A has strategies A_1 and A_2, and player B has strategies B_1 and B_2. Let $X = (x, 1 - x)$ be an arbitrary mixed strategy for A, with $0 \leq x \leq 1$, and let $Y = (y, 1 - y)$ be an arbitrary mixed strategy for B, with $0 \leq y \leq 1$. Let $P_A(x,y)$ stand for the expected pay-off to A when A and B play strategies X and Y, respectively, and let $P_B(x,y)$ be the corresponding expected pay-off to B. Then

$$P_A(x,y) = x(2y + 5(1 - y)) + (1 - x)(3y + (1 - y)) = x(5 - 3y) + (1 - x)(2y + 1)$$
$$= -5xy + 4x + 2y + 1 \quad \text{and}$$

$$P_B(x,y) = x(5y + 2(1 - y)) + (1 - x)(3(1 - y)) = x(3y + 2) + (1 - x)(3 - 3y) = 6xy - x + 3 - 3y$$

We wish to find a Nash equilibrium (X^*, Y^*), where $X^* = (x^*, 1 - x^*)$ and $Y^* = (y^*, 1 - y^*)$. So, first of all, the pay-off $P_A(X^*, Y^*)$ has to be the maximum over all pay-offs $P_A(X, Y^*)$.[4] Thus, for the fixed value y^*, we want to maximize $P_A(x, y^*)$. At a maximum inside the interval $0 \leq x \leq 1$, the (partial) derivative $\partial P_A / \partial x$ must be 0. (The special case when the maximum occurs at an endpoint $x = 0$ or $x = 1$ can be treated separately.) In our example, we get $\partial P_A / \partial x = -5y^* + 4 = 0$. Hence, $y^* = 4/5$. Similarly, the pay-off $P_B(X^*, Y^*)$ has to be the maximum over all pay-offs $P_B(X^*, Y)$. Thus, for the fixed value x^*, we want to maximize $P_B(x^*, y)$. At a maximum inside the interval $0 \leq y \leq 1$, the (partial) derivative $\partial P_B / \partial y$ must be 0. In our example, we get $\partial P_B / \partial y = 6x^* - 3 = 0$. Hence, $x^* = 1/2$. Thus, $X^* = (1/2, 1/2)$ and $Y^* = (4/5, 1/5)$. The corresponding

---

[4]Note that we are using P_A and P_B in two senses: first, as a function of the mixed strategies X and Y, and second, as a function of the numbers x and y that determine X and Y. This should cause no confusion.

pay-offs are $P_A(\frac{1}{2}, 4/5) = 13/5$ and $P_B(\frac{1}{2}, 4/5) = 5/2$. (Let us verify that (X^*, Y^*) actually is a Nash equilibrium. First, a direct computation yields $P_A(x, y^*) = P_A(x, 4/5) = 13/5$. Hence, the maximum of $P_A(x, y^*)$ on the interval $0 \leq x \leq 1$ occurs at every point of the interval and, in particular, at $x = x^* = \frac{1}{2}$. Second, $P_B(x^*, y) = P_B(\frac{1}{2}, y) = 5/2$. So, the maximum of $P_B(x^*, y)$ on the interval $0 \leq y \leq 1$ occurs at every point of the interval and, in particular, at $y = y^* = 4/5$.)

4.3   Graphical method for finding Nash equilibria for 2 × 2 matrices

There is a general graphical method for finding the mixed Nash equilibria for a 2×2 matrix. First note the following simple facts, on which the method is based.

Consider a linear function $f(x) = mx + c$.

1. If $m < 0$, then the maximum of $f(x)$ on the interval $0 \leq x \leq 1$ is attained at $x = 0$.
2. If $m = 0$, then $f(x)$ is constant and the maximum of $f(x)$ on the interval $0 \leq x \leq 1$ is attained at all points of that interval.
3. If $m > 0$, then the maximum of $f(x)$ on the interval $0 \leq x \leq 1$ is attained at $x = 1$.

Now consider any 2×2 matrix

$$\begin{pmatrix} (a_1, b_1) & (a_2, b_2) \\ (a_3, b_3) & (a_4, b_4) \end{pmatrix}$$

As usual, we let $X = (x, 1 - x)$ and $Y = (y, 1 - y)$ be arbitrary mixed strategies for players A and B, respectively. Then the pay-offs when A and B play X and Y are:

$$P_A = x(a_1 y + a_2(1 - y)) + (1 - x)(a_3 y + a_4(1 - y))$$

$$= (a_1 - a_2 - a_3 + a_4)xy + (a_2 - a_4)x + (a_3 - a_4)y + a_4$$

$$= [(a_1 - a_2 - a_3 + a_4)y + (a_2 - a_4)]x + [(a_3 - a_4)y + a_4].$$

$$P_B = x(b_1 y + b_2(1 - y)) + (1 - x)(b_3 y + b_4(1 - y))$$

$$= (b_1 - b_2 - b_3 + b_4)xy + (b_2 - b_4)x + (b_3 - b_4)y + b_4$$

$$= [(b_1 - b_2 - b_3 + b_4)x + (b_3 - b_4)]y + [(b_2 - b_4)x + b_4]$$

Let us draw the following graph G_A. For any y' such that $0 \le y' \le 1$, $P_A(x,y')$ will have the form of a linear function $f(x) = mx + c$ and its maximum value(s) for arguments x on $0 \le x \le 1$ will occur at certain value(s) x' that can be calculated in accordance with (1) to (3) above. G_A will consist of all the points (x',y') obtained in this way. Now assume that $X^* = (x^*, 1 - x^*)$ and $Y^* = (y^*, 1 - y^*)$ form a Nash equilibrium. Then, by the definition of "Nash equilibrium", it is easy to show that the point (x^*, y^*) will lie on the graph G_A. (The reader should carry through this argument in detail.) Similarly, draw the following graph G_B. For any x' such that $0 \le x' \le 1$, $P_B(x',y)$ will have the form of a linear function $f(y) = my + c$ and its maximum value(s) for arguments y on $0 \le y \le 1$ will occur at certain value(s) y' that can be calculated in accordance with (1) to (3) above (with x and y interchanged). G_B is defined to consist of all the points (x',y') obtained in this way. Then, by the definition of "Nash equilibrium," the point (x^*, y^*) will lie on the graph G_B. Thus, all Nash equilibria will lie on the intersection of the graphs G_A and G_B. Conversely, from the construction of G_A and G_B it is easy to see that any point on that intersection will be a Nash equilibrium.

Example 4.7
Consider the game matrix

		y B_1	$1 - y$ B_2
x	A_1	(1,–1)	(2,3)
$1 - x$	A_2	(0,2)	(3,1)

There is no pure Nash equilibrium. Let us use the graphical method for finding Nash equilibria.

$$P_A(x,y) = x(y + 2(1 - y)) + (1 - x)(3(1 - y)) = x(2 - y) + (1 - x)(3 - 3y)$$

$$= 2xy - x + 3 - 3y = (2y - 1)x + (3 - 3y)$$

Consider any y such that $0 \le y \le 1$. When $2y - 1 < 0$, that is, when $y < \frac{1}{2}$, the maximum value for P_A occurs at $x = 0$. Hence, $(0,y)$ lies on G_A. When $2y - 1 = 0$, that is, when $y = \frac{1}{2}$, all x on $0 \le x \le 1$ yield a maximum for P_A. So, all such $(x, \frac{1}{2})$ lie on G_A. When $2y - 1 > 0$, that is, when $y > \frac{1}{2}$, the maximum value for P_A occurs at $x = 1$, and $(1,y)$ would lie on G_A. The graph G_A is shown in Figure 4.1(a) as broken line OUVW. Now let us find G_B.

$$P_B(x,y) = x(-y + 3(1 - y)) + (1 - x)(2y + 1 - y) = x(3 - 4y) + (1 - x)(y + 1)$$

$$= -5xy + 2x + y + 1 = (-5x + 1)y + (2x + 1)$$

Consider any x such that $0 \le x \le 1$. When $-5x + 1 < 0$, that is, when $x > 1/5$, the maximum value for P_B occurs at $y = 0$. Hence, $(x,0)$ lies on G_B. When $-5x + 1 = 0$, that is, when $x = 1/5$, all y on $0 \le y \le 1$ yield a maximum for P_B. So, all such $(1/5,y)$ lie on G_B. When $-5x + 1 > 0$, that is, when $x < 1/5$, the

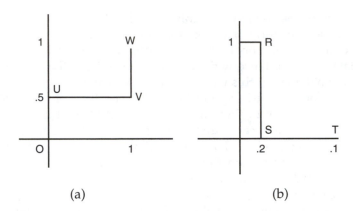

Figure 4.1

maximum value for P_B occurs at $y = 1$, and $(x,1)$ would lie on G_B. The graph G_B is shown in Figure 4.1(b) as broken line $1RST$. The intersection of G_A and G_B is $(1/5,1/2)$. Thus, the unique Nash equilibrium is

$$X^* = (1/5,4/5), \quad Y^* = (1/2,1/2)$$

To verify that (X^*,Y^*) is a Nash equilibrium, observe that

$$P_A(X,Y^*) = (2(^1\!/_2) - 1)x + (3 - 3(^1\!/_2)) = 3/2 = P_A(X^*,Y^*) \text{ and}$$

$$P_B(X^*,Y) = (-5(1/5) + 1)y + 2(1/5) + 1 = 7/5 = P_B(X^*,Y^*)$$

Note that the pay-offs are $(3/2,7/5)$, and these are not as good as the pay-offs $(2,3)$ for (A_1,B_2).

Exercise 4.10
Use both methods (our graphical method and partial derivatives) to find the Nash equilibria and the corresponding pay-offs for the following matrix games.

(a) $\begin{pmatrix} (1,4) & (2,1) \\ (3,0) & (1,2) \end{pmatrix}$ (b) $\begin{pmatrix} (3,2) & (4,3) \\ (1,4) & (6,2) \end{pmatrix}$ (c) $\begin{pmatrix} (2,-1) & (1,0) \\ (1,3) & (2,-2) \end{pmatrix}$

(d) $\begin{pmatrix} (1,0) & (0,1) \\ (2,4) & (3,2) \\ (3,-1) & (2,2) \end{pmatrix}$ (e) $\begin{pmatrix} (3,0) & (2,1) & (1,2) \\ (2,2) & (1,3) & (2,0) \\ (1,1) & (1,0) & (0,1) \end{pmatrix}$ (f) $\begin{pmatrix} (2,4) & (4,3) \\ (1,2) & (0,1) \end{pmatrix}$

(g) $\begin{pmatrix} (2,4) & (1,3) \\ (1,0) & (4,1) \end{pmatrix}$ (h) $\begin{pmatrix} (1,2) & (2,2) \\ (1,3) & (1,0) \end{pmatrix}$ (i) $\begin{pmatrix} (2,1) & (3,1) \\ (2,0) & (3,0) \end{pmatrix}$

Example 4.8

Let us consider the following game matrix.

		y B_1	$1-y$ B_2
x	A_1	(2,1)	(4,0)
$1-x$	A_2	(1,–1)	(5,2)

There are two pure Nash equilibria, (A_1, B_1), yielding (2,1), and (A_2, B_2), yielding (5,2). Let us see whether there is also a mixed equilibrium.

$$P_A(x,y) = x(2y + 4(1 - y)) + (1 - x)(y + 5(1 - y)) = x(4 - 2y) + (1 - x)(5 - 4y)$$

$$= 2xy - x + 5 - 4y$$

$$P_B(x,y) = x(y) + (1 - x)(-y + 2(1 - y)) = xy + (1 - x)(2 - 3y) = 4xy - 2x + 2 - 3y$$

Noting that $P_A(x,y) = (2y - 1)x + 5 - 4y$, we get the three cases:

1. $y < \frac{1}{2}, x = 0$
2. $y = \frac{1}{2}, 0 \le x \le 1$
3. $y > \frac{1}{2}, x = 1$.

Thus, G_A is the graph shown as broken line *OUVW* in Figure 4.2(a). Since $P_B(x,y) = (4x - 3)y - 2x + 2$, we get the three cases:

1. $x < \frac{3}{4}, y = 0,$
2. $x = \frac{3}{4}, 0 \le y \le 1,$ and
3. $x > \frac{3}{4}, y = 1,$
   and the graph G_B is shown as the broken line *OTZW* in Figure 4.2(b).

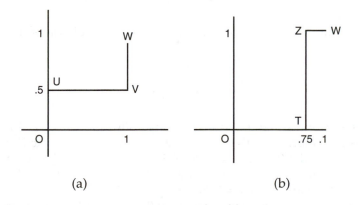

(a)   (b)

Figure 4.2

There are three intersections of the graphs G_A and G_B. The first, at point O, is $x = 0$, $y = 0$, and corresponds to the pure Nash equilibrium (A_2, B_2). The second occurs at $x = .75$, $y = .5$; this is a mixed Nash equilibrium. The third, at point W, is $x = 1$, $y = 1$, and corresponds to the pure Nash equilibrium (A_1, B_1). Thus, a matrix can have both pure and non-pure Nash equilibria. The pay-offs for the mixed Nash equilibrium are $(3, \frac{1}{2})$.

Example 4.9
Let us consider the following game matrix.

$$
\begin{array}{ccc}
 & y & 1-y \\
 & B_1 & B_2 \\
x \quad A_1 & (2,5) & (5,2) \\
1-x \quad A_2 & (1,0) & (0,0)
\end{array}
$$

There is one pure Nash equilibrium (A_1, B_1), yielding $(2,5)$. Let us now check for mixed equilibria.

$$P_A(x,y) = x(2y + 5(1-y)) + (1-x)(y) = x(5-3y) + (1-x)y$$

$$= -4xy + 5x + y$$

$$P_B(x,y) = x(5y + 2(1-y)) + (1-x)(0) = x(3y+2) = 3xy + 2x$$

Since $P_A(x,y) = (-4y+5)x + y$, we have only one case: $y < 5/4$, $x = 1$. (Note that $y \geq 5/4$ is impossible.) Hence, the graph G_A consists of the line segment $x = 1$, $0 \leq y \leq 1$. On the other hand, $P_B(x,y) = 3xy + 2x$. If $x = 0$, all y in $0 \leq y \leq 1$ maximize P_B. If $x > 0$, then we must have $y = 1$ to maximize P_B. So, G_B consists of the line segment from $(0,0)$ to $(0,1)$, plus the line segment from $(0,1)$ to $(1,1)$. Hence, the intersection of G_A and G_B consists of the single point $(1,1)$, that is, $x = 1$, $y = 1$. So, the pure Nash equilibrium (A_1, B_1) is the unique Nash equilibrium.

Example 4.10
Let us use our graphical method to find the Nash equilibria of the matrix

$$
\begin{pmatrix}
(3,1) & (1,4) \\
(1,0) & (1,0)
\end{pmatrix}
$$

(A_1, B_2) is a pure Nash equilibrium with pay-offs $(1,4)$, and (A_2, B_2) is also a pure Nash equilibrium with pay-offs $(1,0)$. Let us look for mixed Nash equilibria. $P_A(x,y) = x(1+2y) + (1-x) = 2yx + 1$. For $y = 0$, P_A attains its maximum at all x such that $0 \leq x \leq 1$. So, the points $(x,0)$ lie on G_A for $0 \leq x \leq 1$. For $0 < y$, P_A attains its maximum at $x = 1$. So, the points $(1,y)$ lie on G_A for $0 < y \leq 1$. Hence, G_A consists of the segment from $(0,0)$ to $(1,0)$, plus the segment from $(1,0)$ to $(1,1)$. Now consider $P_B(x,y) = x(4-3y) = -3xy + 4x$. For $x = 0$, P_B

attains its maximum for all y such that $0 \le y \le 1$. Hence, all points $(0,y)$ with $0 \le y \le 1$ lie on G_B. For $x > 0$, P_B has its maximum at $y = 0$. So, all $(x,0)$ with $0 < x \le 1$ lie on G_B. Thus, G_B consists of the segment from $(0,1)$ to $(0,0)$, plus the segment from $(0,0)$ to $(1,0)$. The intersection of G_A and G_B consists of the segment from $(0,0)$ to $(1,0)$. So, the mixed Nash equilibria are the pairs $X = (x,1 - x)$, $Y = (0,1)$, where $0 < x < 1$. (The pure Nash equilibria are $X = (x,1 - x)$ and $Y = (0,1)$ with $x = 0$ or $x = 1$.) The corresponding pay-offs are $(1,4x)$. This is an example where there are infinitely many Nash equilibria.

Exercise 4.11
Find pure and mixed Nash equilibria and the corresponding pay-offs for the following matrix games:

(a) $\begin{pmatrix} (3,3) & (-1,0) \\ (0,-1) & (0,0) \end{pmatrix}$
(b) $\begin{pmatrix} (0,3) & (0,0) \\ (1,-1) & (-1,0) \end{pmatrix}$

(c) $\begin{pmatrix} (1,2) & (1,3) \\ (1,4) & (1,0) \end{pmatrix}$
(d) $\begin{pmatrix} (1,2) & (1,2) \\ (1,2) & (1,2) \end{pmatrix}$

(e) $\begin{pmatrix} (2,2) & (1,3) & (0,4) \\ (1,3) & (0,4) & (3,1) \\ (2,2) & (1,3) & (2,2) \\ (0,4) & (3,1) & (1,3) \end{pmatrix}$ (Hint: First use dominance relations.)

(f) $\begin{pmatrix} (3,1) & (2,2) \\ (1,2) & (3,0) \end{pmatrix}$
(g) $\begin{pmatrix} (3,0) & (2,1) & (1,2) \\ (2,2) & (1,3) & (2,0) \\ (1,1) & (1,0) & (0,1) \end{pmatrix}$

(h) $\begin{pmatrix} (3,1) & (1,0) \\ (\tfrac{1}{2},2) & (4,1) \end{pmatrix}$
(i) $\begin{pmatrix} (5,3) & (-1,2) \\ (2,0) & (1,1) \end{pmatrix}$

Exercise 4.12
Make a catalogue of all possible types of graphs G_A and G_B for two-person 2×2 games.

Exercise 4.13
Show that there is no two-person 2×2 matrix game with exactly one pure Nash equilibrium and exactly one non-pure mixed Nash equilibrium.

For two-person games in which at least one of the players has more than two strategies, there are algorithms for finding the Nash equilibria, but they are usually too complex to be done by hand and, therefore require computer implementation. One source is the Gambit site, http://econweb.tamu.edu/gambit/.[5]

Let us try to find Nash equilibria for a few low-dimensional game matrices.

Example 4.11

Consider the following matrix, where $(x_1, x_2, 1 - x_1 - x_2)$ and $(y_1, y_2, 1 - y_1 - y_2)$ are arbitrary mixed strategies for players A and B, respectively.

		y_1 B_1	y_2 B_2	$1 - y_1 - y_2$ B_3
x_1	A_1	(2,1)	(2,1)	(0,3)
x_2	A_2	(2,1)	(0,2)	(2,1)
$1 - x_1 - x_2$	A_3	(1,2)	(2,1)	(2,0)

Note that there is no pure Nash equilibrium. Let P_A and P_B be the pay-off functions for A and B. Then:

$$P_A = x_1(2y_1 + 2y_2) + x_2(2 - 2y_2) + (1 - x_1 - x_2)(2 - y_1)$$

$$= 3x_1y_1 + 2x_1y_2 + x_2y_1 - 2x_2y_2 - 2x_1 + 2 - y_1.$$

$$P_B = x_1(3 - 2y_1 - 2y_2) + x_2(y_2 + 1) + (1 - x_1 - x_2)(2y_1 + y_2)$$

$$= -4x_1y_1 - 3x_1y_2 - 2x_2y_1 + 3x_1 + x_2 + 2y_1 + y_2$$

Assume that $X^* = (x_1^*, x_2^*, 1 - x_1^* - x_2^*)$ and $Y^* = (y_1^*, y_2^*, 1 - y_1^* - y_2^*)$ form a Nash equilibrium. Now, $P_A(X, Y^*)$ should have a maximum at $X = X^*$. So, the partial derivatives $\partial P_A/\partial x_1$ and $\partial P_A/\partial x_2$ should be 0. (Exceptions might appear if the maximum occurs at an endpoint.) So, $\partial P_A/\partial x_1 = 3y_1^* + 2y_2^* - 2 = 0$ and $\partial P_A/\partial x_2 = y_1^* - 2y_2^* = 0$.

Solving, we obtain $y_1^* = {}^1/_2$ and $y_2^* = {}^1/_4$. Hence, $Y^* = ({}^1/_2, {}^1/_4, {}^1/_4)$. Similarly, $P_B(X^*, Y)$ should have a maximum at $Y = Y^*$. So, the partial derivatives $\partial P_B/\partial y_1$ and $\partial P_B/\partial y_2$ should be 0. Then $\partial P_B/\partial y_1 = -4x_1^* - 2x_2^* + 2 = 0$ and $\partial P_B/\partial y_2 = -3x_1 + 1 = 0$. Solving, we get $x_1^* = {}^1/_3$ and $x_2^* = {}^1/_3$. Hence, $X^* = ({}^1/_3, {}^1/_3, {}^1/_3)$. We must check that (X^*, Y^*) is an equilibrium pair. First, $P_A(X, Y^*) = 3/2\ x_1 + {}^1/_2\ x_1 + {}^1/_2\ x_2 - {}^1/_2\ x_2 - 2x_1 + 2 - {}^1/_2 = 3/2$. So, X^* maximizes $P_A(X, Y^*)$. Second, $P_B(X^*, Y) = -4/3\ y_1 - y_2 - 2/3\ y_1 + 1 + 1/3 + 2y_1 + y_2 = 4/3$. Hence, Y^* maximizes $P_B(X^*, Y)$. Thus, (X^*, Y^*) actually is a Nash equilibrium. Note also that the corresponding pay-offs are (3/2, 4/3).

---

[5]For a general survey, see McKelvey-McLennan [1996]. Algorithms for obtaining one Nash equilibrium for a two-person game are discussed in Lemke-Howson [1964] and Eaves [1971]. For obtaining all Nash equilibria, see Mangasarian [1964] and Dickhaut-Kaplan [1991] For extensions to n-person games, consult Rosenmüller [1971] and Wilson [1971].

Exercise 4.14

Find Nash equilibria and corresponding pay-offs for the following game matrices.

$$\text{(a)} \begin{pmatrix} (5,1) & (2,2) & (1,3) \\ (-1,2) & (3,1) & (2,1) \\ (0,3) & (1,3) & (2,2) \end{pmatrix} \qquad \text{(b)} \begin{pmatrix} (0,2) & (3,-1) & (1,1) \\ (6,0) & (1,2) & (3,3) \\ (3,2) & (2,3) & (4,1) \end{pmatrix}$$

$$\text{(c)} \begin{pmatrix} (2,1) & (2,1) & (0,3) \\ (2,1) & (0,3) & (2,1) \\ (1,2) & (2,1) & (2,1) \end{pmatrix} \qquad \text{(d)} \begin{pmatrix} (5,2) & (3,4) & (-2,3) \\ (3,2) & (1,2) & (4,2) \\ (0,4) & (2,2) & (1,5) \end{pmatrix}$$

$$\text{(e)} \begin{pmatrix} (3,1) & (5,2) & (4,3) \\ (2,2) & (4,1) & (5,0) \\ (1,1) & (3,2) & (3,1) \end{pmatrix} \qquad \text{(f)} \begin{pmatrix} (1,4) & (0,0) & (0,0) & (0,0) \\ (0,0) & (2,3) & (0,0) & (0,0) \\ (0,0) & (0,0) & (3,2) & (0,0) \\ (0,0) & (0,0) & (0,0) & (4,1) \end{pmatrix}$$

Example 4.12

Let us look at the following 3×2 game matrix.

		y B_1	$1-y$ B_2
x_1	A_1	(3,1)	(1,3)
x_2	A_2	(2,2)	(2,1)
$1 - x_1 - x_2$	A_3	(1,1)	(3,0)

The pay-off functions are:

$$P_A = x_1(2y + 1) + x_2(2) + (1 - x_1 - x_2)(3 - 2y) = 4x_1y + 2x_2y - 2x_1 - x_2 + 3 - 3y.$$

$$P_B = x_1(3 - 2y) + x_2(y + 1) + (1 - x_1 - x_2)y = -3x_1y + 3x_1 + x_2 + y$$

Assume $X^* = (x_1{}^*, x_2{}^*, 1 - x_1{}^* - x_2{}^*)$ and $Y^* = (y^*, 1 - y^*)$ form a Nash equilibrium. $P_A(X, Y^*)$ is maximized at $X = X^*$. So, let's see what happens when the partial derivatives with respect to x_1 and x_2 are set equal to 0.

$$\partial P_A/\partial x_1 = 4y - 2 = 0 \text{ and } \partial P_A/\partial x_2 = 2y - 1 = 0$$

In each case, we get $y = {}^1/_2$. Similarly, let $\partial P_B/\partial y = -3x_1 + 1 = 0$, so that $x_1 = 1/3$. X^* is not completely determined. It has the form $X^* = (1/3, a, 2/3 - a)$, with $0 \le a \le 2/3$. Let us check that (X^*, Y^*) form a Nash equilibrium. First, $P_A(X, Y^*) = 2x_1 + x_2 - 2x_1 - x_2 + 2 = 2$, so that $P_A(X, Y^*)$ is maximal at $X = X^*$. Second, $P_B(X^*, Y) = -y + 1 + a + y = a + 1 = P_B(X^*, Y^*)$. Thus, there are infinitely many Nash equilibria. The pay-offs for (X^*, Y^*) are $(2, a + 1)$.

The methods involving partial derivatives that were used in Examples 4.11 and 4.12 don't always work. The reader should try to apply them to the matrices

$$\begin{pmatrix} (2,1) & (4,0) & (-1,2) \\ (3,2) & (1,3) & (2,0) \end{pmatrix}$$

and

$$\begin{pmatrix} (1,4) & (3,1) & (2,5) \\ (0,2) & (-2,3) & (3,1) \\ (2,1) & (1,2) & (0,2) \end{pmatrix}$$

Example 4.13
Consider the following game matrix.

$$\begin{pmatrix} (1,0) & (2,5) & (0,4) \\ (0,3) & (4,1) & (2,2) \end{pmatrix}$$

Then $P_A(x,y) = x(y_1 + 2y_2) + (1-x)(2 - 2y_1 + 2y_2) = 3xy_1 - 2x + 2 - 2y_1 - 2y_2$ and

$P_B(x,y) = x(4 - 4y_1 + y_2) + (1-x)(2 + y_1 - y_2) = -5xy_1 + 2xy_2 + 2x + 2 + y_1 - y_2$

So, $\partial P_A/\partial x = 3y_1 - 2 = 0$ yields $y_1 = 2/3$. Also, (*) $\partial P_B/\partial y_1 = -5x + 1 = 0$ yields $x = 1/5$, and (**) $\partial P_B/\partial y_2 = 2x - 1 = 0$ yields $x = {}^1/_2$. (*) and (**) are incompatible. So, let us first assume (*). Then $\partial P_B/\partial y_2 = -3/5 < 0$, P_B would decrease with respect to y_2 and, therefore, P_B would assume its maximum at $y_2 = 0$. This leads us to the pair $X^* = (1/5, 4/5)$ and $Y^* = (2/3,0,1/3)$. Let us check that they form an equilibrium pair. First, $P_A(X,Y^*) = 2/3$, so that $P_A(X,Y^*)$ achieves a maximum at $X = X^*$. Second, $P_B(X^*,Y) = 12/5 - y_2 \le 12/5 = P_B(X^*,Y^*)$, so that $P_B(X^*,Y)$ has a maximum at $Y = Y^*$. (The reader should verify that the case (**) does not lead to an equilibrium pair.)

Exercise 4.15
Find (pure and mixed) Nash equilibria and the corresponding pay-offs for the following matrices:

$$\text{(a)} \begin{pmatrix} (2,1) & (0,0) & (1,2) \\ (1,2) & (2,1) & (0,0) \\ (0,0) & (1,2) & (2,1) \end{pmatrix} \qquad \text{(b)} \begin{pmatrix} (5,1) & (2,2) & (1,3) \\ (-1,2) & (3,1) & (2,1) \\ (0,3) & (1,3) & (2,2) \end{pmatrix}$$

$$(c) \begin{pmatrix} (2,0) & (0,2) \\ (1,1) & (1,1) \\ (0,2) & (2,0) \end{pmatrix} \qquad (d) \begin{pmatrix} (0,2) & (2,1) & (3,-1) \\ (3,1) & (0,2) & (1,4) \end{pmatrix}$$

$$(e) \begin{pmatrix} (6,-1) & (2,5) \\ (4,3) & (3,1) \\ (3,5) & (4,0) \end{pmatrix} \qquad (f) \begin{pmatrix} (3,7) & (1,2) & (4,1) \\ (2,3) & (3,4) & (5,2) \\ (4,1) & (5,1) & (3,5) \end{pmatrix}$$

$$(g) \begin{pmatrix} (2,0) & (-3,4) & (3,1) \\ (1,3) & (0,2) & (4,2) \\ (2,1) & (3,0) & (2,4) \end{pmatrix} \qquad (h) \begin{pmatrix} (1,2) & (4,0) & (3,4) \\ (2,2) & (1,3) & (4,1) \end{pmatrix}$$

$$(i) \begin{pmatrix} (1,1) & (-1,3) & (2,2) \\ (4,5) & (5,0) & (1,3) \end{pmatrix} \qquad (j) \begin{pmatrix} (2,5) & (-1,3) & (4,1) \\ (1,2) & (0,4) & (3,5) \end{pmatrix}$$

$$(k) \begin{pmatrix} (1,1) & (1,1) & (3,0) \\ (1,1) & (0,0) & (2,0) \end{pmatrix} \qquad (l) \begin{pmatrix} (1,1) & (0,0) \\ (0,0) & (0,0) \\ (-1,-1) & (0,-2) \end{pmatrix}$$

$$(m) \begin{pmatrix} (2,1) & (4,0) & (-1,2) \\ (3,2) & (1,3) & (2,0) \end{pmatrix}$$

Let us now look at some interesting examples of two-person non-zero-sum games.

Example 4.14
(Threats and Cooperative Games) Consider the game

	B_1	B_2
A_1	(3,2)	(2,1)
A_2	(2,-1000)	(1,-2000)

Here, (A_1, B_1) is the unique pure Nash equilibrium, with pay-offs (3,2). In fact, (A_1, B_1) is also the best pair. Moreover, it is easy to check that there are no other mixed Nash equilibria. So, from our standard game-theoretic viewpoint, nothing more is to be said. However, notice that player B would suffer devastating losses if player A plays strategy A_2. A's pay-off would go down from 3 to only 2 or 1, but B's loss would be at least 1000. (B would probably choose strategy B_2, and A's pay-off would decrease to 2.) Hence, A can threaten to play strategy A_2 unless B pays A a suitable "bribe" not to do so.[6] Of course, we have no provisions for such threats and bribes, but, if game theory is to be applied to real-life situations, we eventually will want to allow for pre-play negotiations and agreements. Sometimes, a bribe is given the more respectable name, "side payment". Even when such payments are permitted, it is not clear how to arrive at an appropriate amount. This is the case in Example 4.14.

Games that allow for pre-play communication, or enforceable pre-play contracts, or side payments are called *cooperative games*. The games that we considered before are called *non-cooperative games*. The adjective "cooperative" does not imply that the players are concerned about each other's welfare. It signifies only that binding agreements can be made about strategies to be played, about division of pay-offs, and about side payments.

Exercise 4.16

Examine the possible role of threats in the games given by the following matrices. (Determine also whether counterthreats are conceivable.)

(a) $\begin{pmatrix} (2,20) & (4,25) \\ (0,3) & (1,2) \end{pmatrix}$ (b) $\begin{pmatrix} (2,20) & (6,25) \\ (0,3) & (1,2) \end{pmatrix}$

(c) $\begin{pmatrix} (20,2) & (0,3) \\ (24,8) & (1,2) \\ (23,9) & (1,6) \end{pmatrix}$ (d) $\begin{pmatrix} (13,-1) & (3,0) \\ (12,2) & (1,5) \\ (8,6) & (2,1) \end{pmatrix}$

(e) $\begin{pmatrix} (3,2) & (2,3) \\ (2,-1000) & (3,-2000) \end{pmatrix}$ (f) $\begin{pmatrix} (5,2) & (2,5) \\ (2,-1000) & (3,-2000) \end{pmatrix}$

(g) The matrix of Example 4.9   (h) $\begin{pmatrix} (3,3) & (0,2) \\ (2,0) & (1,1) \end{pmatrix}$

---

[6]We are assuming that A and B attribute comparable values to their pay-off numbers, so that a loss of 1000 to B is much more significant than a loss of 1 to A.

Example 4.15

(Pre-play Commitments) It seems reasonable to assume that knowledge in advance of an opponent's move might very well help, and certainly not hurt, the player who gets the extra information. But this "obvious fact" turns out not to be always reliable, as the following cases show:

1. If a player announces in advance a binding commitment to play a certain strategy, then this should not give him any advantage and might very well help his opponent. However, consider the following game:

	B_1	B_2
A_1	(3,1)	(0,0)
A_2	(0,0)	(1,3)

(A_1,B_1) and (A_2,B_2) are pure Nash equilibria, but player A would prefer the first and player B the second. A standard computation shows that there is a mixed Nash equilibrium, $X^* = (3/4,1/4)$ and $Y^* = (1/4,3/4)$, with pay-offs $(3/4,3/4)$. If player A announced beforehand a binding commitment to play strategy A_1, then player B would feel compelled to play strategy B_1 (since, otherwise, if he played strategy B_2, he would get a pay-off of 0 instead of 1). Hence, A's pre-play commitment yields a pay-off of 3 to player A and 1 to player B. That would be advantageous to A, but not to B. Of course, if player B had had the foresight to make a pre-play commitment to play strategy B_2 before A made a commitment, then B would have gained the advantage. Thus, for the sake of fairness, a binding commitment by one player would have to be acceptable to the other player. We shall see later that there is a more profitable way for the players to behave in a cooperative game based on the matrix in this and other examples.

2. **(Foreknowledge)** Assume that the rules of a game are modified so that player A first chooses his strategy, B is informed of A's choice, and then B is allowed to choose his strategy. It seems clear that this should work to B's advantage. Consider the game

	B_1	B_2	B_3
A_1	(4,1)	(7,0)	(1,5)
A_2	(5,2)	(6,3)	(0,5)
A_3	(6,4)	(5,7)	(−1,3)

B finds out A's move before B has to move. Then A can reason as follows:
If I play A_1, B will choose B_3, and the outcome will be (1,2). If I play A_2, B will choose B_2, and the outcome will be (6,3). If I play A_3, B will choose B_2, and the outcome will be (5,7). Hence, A will choose A_2

and the result will be (6,3), a rather good result for A, but a mediocre outcome for B. (Note that (A_1,B_3) is the only pure Nash equilibrium, yielding pay-offs (1,5).) Similarly, if A had foreknowledge of B's strategy, the result would be (1,5), bad for A, and reasonably good for B. Such counterintuitive results would not appear in many other cases. The reader can see what would happen in Exercises 4.2(b), 4.11(e), 4.15(g), 4.16(a to c).

Example 4.16

Prisoner's Dilemma (PD)[7] A pair of criminals A and B have been arrested for a crime and are being interrogated separately. The district attorney offers them the following alternatives. If they both confess, then each will receive a four-year sentence. If one of them confesses and the other remains silent, then the one who confesses will receive a one-year sentence, while the obdurate one will be prosecuted to the full extent of the law and receive an eight-year sentence. If they both remain silent, their lawyers assure them that they will get away with only a two-year sentence. Each of A and B has two strategies, Confess (C) or Remain Silent (S). Here is the game matrix.

	C_B	S_B
C_A	(−4,−4)	(−1,−8)
S_A	(−8,−1)	(−2,−2)

The separate pay-off matrices for A and B are:

	C_B	S_B
C_A	− 4	−1
S_A	− 8	−2

A

	C_B	S_B
C_A	−4	−8
S_A	−1	−2

B

For player A, the first row C_A dominates the second row S_A and, therefore, A will choose strategy C_A. Likewise, for player B, the first column C_B dominates the second column S_B, and, therefore, B will play strategy C_B. Thus, both A and B will confess. So, the pay-offs are (−4,−4), that is, both will receive sentences of four years. Notice that the chosen pair of strategies (C_A,C_B) is the only Nash equilibrium. On ordinary game-theoretic grounds, confessing would be advisable for the players. However, the pay-offs (−4,−4) are clearly not the best result that can be achieved. If both A and B remain silent, then this pair of strategies (S_A,S_B) yields the outcome (−2,−2), that is, two-year sentences for A and B. Note that (S_A,S_B) is not a Nash equilibrium,

---

[7]This example was formulated and named by Albert Tucker and was based on work of Merrill Flood and Melvin Dresher at the Rand Corporation in 1950.

but it is Pareto-optimal. However, it would be extremely dangerous for A or B to remain silent. For, if one of them remains silent, then the other might betray him by confessing. In that case, the confessor would receive the light sentence of one year while the silent one would get the hard sentence of eight years. Both A and B realize that it would be better for them to remain silent, but neither can take that risk. They are not allowed to confer in advance to map out their strategies. But even if they were allowed to do so and they promised each other that they would remain silent, there is no way that such a promise can be enforced and they still run the risk of being betrayed. (Later on, we shall consider games in which certain kinds of enforceable contracts can be made in advance.) In general, by a Prisoner Dilemma game we mean any two-person game:

$$
\begin{array}{ccc}
 & B_1 & B_2 \\
A_1 & (a,a) & (c,d) \\
A_2 & (d,c) & (b,b)
\end{array}
$$

where $d < a < b < c$. In interesting cases, b is significantly bigger than a, and d is very much smaller than a.[8] In the example above, $d = -8$, $a = -4$, $b = -2$, and $c = -1$. Other examples of Prisoner Dilemma games are

$$
\begin{pmatrix}
(0,0) & (5,-5) \\
(-5,5) & (2,2)
\end{pmatrix}
$$

and

$$
\begin{pmatrix}
(3,3) & (11,1) \\
(1,11) & (8,8)
\end{pmatrix}
$$

It is easy to see that (A_1, B_1) is the only Nash equilibrium, (A_2, B_2) is Pareto-optimal, and (A_2, B_2) is preferable to (A_1, B_1) for both players. $((A_1, B_2)$ and (A_2, B_1) are also Pareto-optimal.) If A and B considered cooperating with each other by playing A_2 and B_2, they run the risk of getting a much lower pay-off if the opponent "defects" by switching to the alternative uncooperative Nash strategy. If only one play of the game is involved, then there seems to be no doubt that the players should defect and play A_1 and B_1.

However, if the game is to be played many times, then it would be extremely profitable for both players to trust each other enough to start playing strategies A_2 and B_2. Of course, this problem is more psychological

---

[8]Sometimes, it is also assumed that $b > \frac{1}{2}(c + d)$, as in Example 4.16. This ensures that, if the game is played again and again, the Pareto-optimal pay-offs (b,b) would be better than what the players can get in the long run by agreeing to alternate strategy pairs (A_1, B_2) and (A_2, B_1).

or political than mathematical.[9] Accordingly, it has received a great deal of attention from social scientists (usually under the name, Iterated Prisoners' Dilemma). Tournaments have been run between long-range strategies for playing the iterated game. There are a host of such strategies, including Always Cooperate, Always Defect, Tit for Tat (i.e., cooperate on the first play and thereafter cooperate when and only when the opponent did not defect on the previous play), and Holding a Grudge (i.e., cooperate until the opponent defects and then always defect). If the competition consists of a fixed number of games, then a kind of backward induction argues for both players always defecting. (In fact, the last game of the competition would be just like a single game of PD and, therefore, both players should defect. Then the next-to-the-last game is essentially like a single game and both players should defect. This argument can be carried all the way back to the first game, with the result that the players should always defect.) So, it seems better not to assume that there is a definite number of games. Much of the original work and analyses of these competitions have been done by the political scientist Robert Axelrod. (See the books by Axelrod [1984, 1997], the paper Axelrod-Dion [1988], and the annotated bibliography for 1988–1994 by Axelrod and Lisa D'Ambrosio on the Internet.) Possible applications to evolutionary biology arise when poorly performing strategies are eliminated and only the surviving strategies are permitted in further competitions.[10]

Exercise 4.17

In a Prisoner Dilemma game, can there be a mixed Nash equilibrium (different from (A_1, B_1), of course)?

Example 4.17

(Chicken) Two hot-rod drivers A and B drive their cars at high speed toward each other in the middle of the road. Each driver can either keep on driving straight ahead, daring the other driver to get out of the way, or the driver can swerve and avoid a collision. We denote these two strategies by D (dare) and S (swerve).

Let us assume that the table of pay-offs below reflects the following facts:

	D_B	S_B
D_A	(−50,−50)	(4,−2)
S_A	(−2,4)	(−2,−2)

[9]An example in international relations is that of an arms race between two countries. Each country has two strategies: spending either an enormous or a moderate amount on its armed forces.
[10]In the first tournament, 14 computer-programmed strategies competed, along with Random (i.e., cooperate half the time at random). Each program played against itself and the other programs for 200 games. (Results of past encounters could be used.) The tournament was run 5 times, and Tit for Tat was the winner. There was a later tournament with 62 competitors. The number of games was indeterminate, instead of being 200. Tit for Tat won again. Also see Nachbar [1992].

In the case of a collision (D_A, D_B), both drivers are killed. If both swerve, they suffer a loss of respect from their friends. If only one swerves, he suffers that same loss, but the driver who didn't swerve is much admired. The numbers actually used are quite arbitrary. (D_A, S_B) and (S_A, D_B) are pure Nash equilibria, but there is no plausible way in which they might arise. The methods we have used earlier show that $X = (1/9, 8/9)$ and $Y = (1/9, 8/9)$ form a mixed Nash equilibrium, with pay-offs $(-2, -2)$, the same as those for (S_A, S_B). Thus, the drivers could swerve with probability 8/9 and achieve the same outcome as if they both always swerved.

Exercise 4.18
Consider the following general "Chicken" game, where $0 < L < M < K$:

$$
\begin{array}{ccc}
 & D_B & S_B \\
D_A & (-K,-K) & (M,-L) \\
S_A & (-L,M) & (-L,-L)
\end{array}
$$

Show the mixed Nash equilibria are

$$X = \left(\frac{L+M}{K+M}, \frac{K-L}{K+M} \right) = Y$$

and the pay-offs are $(-L, -L)$.

Example 4.18
(Battle of the Sexes) Assume that a husband (H) and wife (W) want to go out on a certain day. The husband prefers a sporting event (S) and the wife would rather go to a crafts exhibit (C). They can go out separately, but they would prefer to be together. Let us assume that the following table is a fair representation of the situation,[11] where S_H and C_H stand for the two possible choices ("strategies") the husband can select, and similarly S_W and C_W for the wife. (The specific numbers are somewhat arbitrary.)

$$
\begin{array}{ccc}
 & S_W & C_W \\
S_H & (6,3) & (4,4) \\
C_H & (0,0) & (3,6)
\end{array}
$$

(S_H, C_W) is the unique Nash equilibrium. (The reader should verify that this includes mixed strategies.) The pay-offs $(4,4)$ indicate that "doing one's own thing" is valued slightly more than the pay-off 3 for accompanying one's spouse at an event that the spouse prefers.

---

[11]This game seems to have been initially discussed and named by Luce-Raiffa [1957]. There are several essentially different ways of choosing the game matrix. Also see Rapoport [1966], pp. 95–99.

Exercise 4.19

Examine what happens in the Battle of the Sexes (Example 4.18) when the relative importance of being with one's spouse is altered in two different ways, as in Tables (a) and (b) below.

(a)

	S_W	C_W
S_H	(6,4)	(3,3)
C_H	(0,0)	(4,6)

(b)

	S_W	C_W
S_H	(6,3)	(3,3)
C_H	(0,0)	(3,6)

In (a), being with one's spouse is valued a little more, 4, than "doing one's own thing," 3, but, in (b), they are valued equally.

Exercise 4.20

1. Study what happens when the Battle of the Sexes (Example 4.18) is generalized to the following game, with $N < L < M < K$.

	S_W	C_W
S_H	(K,L)	(M,M)
C_H	(N,N)	(L,K)

2. Examine what happens when the two variants of the Battle of the Sexes in Exercise 4.19 are generalized in a manner similar to that of part (1) of this exercise.

3. Sometimes the Battle of the Sexes is taken in the following different forms:

    (a)

	S_W	C_W
S_H	(2,1)	(0,0)
C_H	(0,0)	(1,2)

    (b)

	S_W	C_W
S_H	(2,1)	(−1,−1)
C_H	(−1,−1)	(1,2)

Analyze these forms of the game. Game (b) was the original game introduced in Luce-Raiffa [1957].

4.4   Inadequacies of Nash equilibria
in non-zero sum games. Cooperative games

We already know that non-zero-sum games always possess at least one Nash equilibrium, possibly a pair of mixed strategies. We also know, however, that these equilibria do not always tell the players how they should play the game. Here are some of the problems that may arise.

1. There may be several equilibria, with different payoffs and with no clear choice between them. An extension to a cooperative game might resolve the problem.
2. The Nash equilibria may not yield the best outcomes. There may be several non-equivalents schemes to suggest procedures for the players to follow. Some require changing to a cooperative game.

Example 4.19
Consider the game

$$\begin{pmatrix} (1,2) & (0,0) \\ (0,0) & (2,1) \end{pmatrix}$$

Both (A_1,B_1) and (A_2,B_2) are pure Nash equilibria. If $(x,1-x)$ and $(y,1-y)$ are mixed strategies for A and B, respectively, then, as before, we can compute the pay-offs P_A and P_B for A and B, respectively:

$$P_A(x,y) = 3xy - 2x + 2 - 2y \quad \text{and} \quad P_B(x,y) = 3xy - x + 1 - y$$

The techniques developed earlier (using either partial derivatives or the graphical method) show that $X^* = (1/3, 2/3)$ and $Y^* = (2/3, 1/3)$ form a Nash equilibrium, with pay-offs $(2/3, 2/3)$. However, these pay-offs are inferior for both players to the pay-offs $(1,2)$ and $(2,1)$ at the pure equilibria. Moreover, since B prefers $(1,2)$ and A prefers $(2,1)$, we seem to have reached an impasse.

Letting $u = P_A(x,y)$ and $v = P_B(x,y)$, look at the set $\mathcal{R}$ of possible pay-offs (u,v) in the uv-plane, as shown in Figure 4.3. This is the range of the function that maps all points (x,y) of the unit square $0 \le x \le 1$, $0 \le y \le 1$ onto the points (u,v) such that

$$u = 3xy - 2x + 2 - 2y, \, v = 3xy - x + 1 - y$$

R is $(1,2)$, S is $(2,1)$, and O is the origin $(0,0)$. The boundary of the unit square is mapped twice onto the segments SO and RO. (For example, the side $y = 0$,

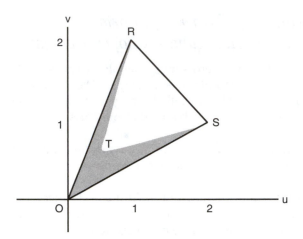

Figure 4.3

$0 \leq x \leq 1$ maps onto *SO*.) The region $\mathcal{R}$ turns out to be a three-sided figure whose sides are the segments *SO* and *RO* and a curve *RTS* that is the image of the diagonal $y = x$. Thus, *RTS* consists of all points (u,v) such that

$$u = 3x^2 - 4x + 2, v = 3x^2 - 2x + 1$$

where $0 \leq x \leq 1$. T is the point $(3/4,3/4)$ corresponding to $x = {}^1/_2$. Note that $\mathcal{R}$ is not a convex set, that is, there are pairs of points in $\mathcal{R}$ for which the segment connecting them does not lie entirely within $\mathcal{R}$. For example, the midpoint $(3/2,3/2)$ of the segment *RS* does not lie in $\mathcal{R}$.

To see this, assume for the sake of contradiction that $3/2 = 3xy - 2x + 2 - 2y$ and $3/2 = 3xy - x + 1 - y$ for some (x,y) in the unit square. By some simple algebra we can deduce that $x = {}^1/_2 (2 \pm \sqrt{-4})$, which is impossible. Thus, $(3/2,3/2)$ is not the pay-off for any pair of mixed strategies. Nevertheless, $(3/2,3/2)$ might be considered a reasonable compromise between the outcomes $(1,2)$ and $(2,1)$, since the pay-offs would be the averages of the pay-offs at $(1,2)$ and $(2,1)$.[12] Moreover, the outcome would be achievable if the players A and B agreed to randomly play with equal probability the strategy pairs (A_1,B_1) and (A_2,B_2). This would require an enforceable cooperation between A and B, so that they would now be playing a cooperative game.

In general, this new cooperative game would permit A and B to play any *joint randomized strategy* (*jrs*), that is, an agreement to play each pair (A_i,B_j) of strategies with a probability z_{ij} such that $\sum_{i,j} z_{ij} = 1$. The pay-offs for such a *jrs* would be $\sum_{i,j} z_{ij} P_A(A_i,B_j)$ and $\sum_{i,j} z_{ij} P_B(A_i,B_j)$ for A and B, respectively. In

---

[12]It is by no means certain that the players would accept this compromise. Economic or psychological factors might enter into the players' calculations.

our example, the pay-off (3/2,3/2) is obtained from the *jrs* that assigns probability $1/2$ to each of the pairs (A_1,B_1) and (A_2,B_2) and probability 0 to the others. In general, if $\mathcal{R}$ denotes the set of possible pay-offs (u,v) of a game, then the set of pay-offs corresponding to all joint randomized strategies will be a region $\mathcal{R}^*$ that includes $\mathcal{R}$. Moreover, $\mathcal{R}^*$ will be convex, since any point on a line segment between two points of $\mathcal{R}^*$ corresponds to a suitable *jrs* obtained from those *jrs*'s that determine the given two points. $\mathcal{R}^*$ will be the smallest convex set containing $\mathcal{R}$, that is, the intersection of all convex sets containing $\mathcal{R}$. In our example, $\mathcal{R}^*$ is the triangle *ORS* in Figure 4.3.

In general, the players A and B will only want to consider outcomes (u,v) in the set $\mathcal{R}^*$ for which there is no *better* outcome (u',v') in the set $\mathcal{R}^*$, that is, an outcome different from (u,v) such that $u \le u'$ and $v \le v'$. Points (u,v) for which there is no such better outcome can be called *Pareto optimal* outcomes. In the graph of $\mathcal{R}^*$, they are points of $\mathcal{R}^*$ having no other points of $\mathcal{R}^*$ to the north-east of them. The set $\mathcal{N}$ of Pareto optimal outcomes is called the *negotiation set* of the game, since one may assume that the players A and B may want to conduct a negotiation as to which outcome in $\mathcal{N}$ should be chosen. (Some authors use the term *bargaining set* instead of *negotiation set*.) In Figure 4.3, the negotiation set is the line segment *RS*.

Example 4.20
Consider the game

$$\begin{pmatrix} (3,1) & (2,2) \\ (0,5) & (3,4) \end{pmatrix}$$

There is no pure Nash equilibrium. The pay-offs are:

$$u = P_A(x,y) = 4xy - x + 3 - 3y \quad \text{and}$$

$$v = P_B(x,y) = -2xy - 2x + y + 4$$

The usual procedure yields the mixed Nash equilibrium $X^* = (1/2,1/2)$, $Y^* = (1/4,3/4)$, with pay-offs (9/4,3). Note that this pay-off is smaller than the pay-offs (3,4) yielded by a non-equilibrium pair (A_2,B_2), so that no Nash equilibrium is helpful in this game. The graph $\mathcal{R}^*$ of pay-off outcomes of joint randomized strategies is the boundary and interior of quadrilateral *RSTU* in Figure 4.4, where *R, S, T, U* are the pay-off pairs (0,5), (3,4), (3,1), (2,2). (Point *n* is the pay-off pair (9/4,3) from the Nash equilibrium.) The negotiation set $\mathcal{N}$ is side *RS*. Without further information about the players, we cannot with certainty say that the midpoint (3/2,9/2) of *RS* is a reasonable compromise.

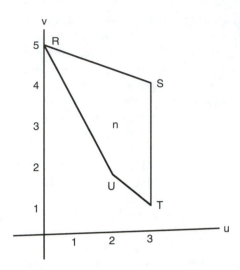

Figure 4.4

Example 4.21

   (a)   Consider the game

$$\begin{pmatrix} (4,3) & (2,5) \\ (1,1) & (5,0) \end{pmatrix}$$

There is no pure Nash equilibrium. The pay-offs are:

$$u = P_A(x,y) = 6xy - 3x + 5 - 4y \quad \text{and}$$

$$v = P_B(x,y) = -3xy + 5x + y$$

The usual procedure yields the mixed Nash equilibrium $X^* = (1/3,2/3)$, $Y^* = (1/2,1/2)$, with pay-offs $(3,5/3)$. Note that this pay-off is smaller than the pay-offs $(4,3)$ yielded by a non-equilibrium pair (A_1,B_1), so that no Nash equilibrium is helpful in this game. The set $\mathcal{R}^*$ of pay-offs of all joint randomized strategies is the boundary and interior of quadrilateral *RSTU* in Figure 4.5. *n* indicates the outcome of the Nash equilibrium. The broken line *RST* is the negotiation set.

   (b)   **(Intersubjective Values. Side Payments)** Consider the game matrix

$$\begin{pmatrix} (7,1) & (1,2) \\ (2,1) & (3,0) \end{pmatrix}$$

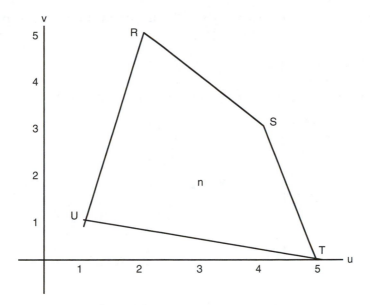

Figure 4.5

There is no Nash equilibrium of pure strategies. The pay-off functions are $u = P_A(x,y) = 7xy - 2x + 3 - y$ and $v = P_B(x,y) = -2xy + 2x + y$. Either of our two methods for finding mixed Nash equilibria yields $X = (1/2, 1/2)$, $Y = (2/7, 5/7)$. The pay-offs at the mixed Nash equilibrium are $(19/7, 1)$. If the graph of pay-offs yielded by joint randomized strategies is drawn, it can be seen that the negotiation set is the line segment connecting $(1,2)$ and $(7,1)$. We cannot say that the midpoint $(4, 3/2)$ of that segment should be agreed upon, unless we know more about the preferences of A and B. If we assume that A and B have the same system of preferences,[13] this still does not determine what should be done, but various possibilities suggest themselves. A and B could play the strategy pair (A_1, B_1) that yields the largest total pay-off, $7 + 1 = 8$. A could then give B a side payment of 3, with the pay-offs $(4,4)$ as the final result. A better scheme might be to divide the total pay-off 8 into two parts that reflect the "relative strengths" of A and B in the original game, but a unique, reasonable way of measuring the relative strengths is hard to agree on.

Exercise 4.21

For each of the following matrix games, find all Nash equilibria and their corresponding pay-offs, describe the set $\mathcal{R}$ of pay-offs for all pairs of mixed strategies, draw the graph $\mathcal{R}^*$ of the pay-offs for all joint randomized

---

[13]With good reason, such an assumption is generally frowned on in the social sciences and game theory, but it is sometimes plausible and often a convenient working hypothesis.

strategies, identify the negotiation set, and weigh the merits of possible side payments:

1. Exercises 4.2(b,c)
2. Examples 4.5 to 4.8
3. Exercises 4.6, 4.7(b to d), 4.10(a-i), 4.11(a-i), 4.19(a,b).

4. $\begin{pmatrix} (1,4) & (4,2) \\ (3,3) & (5,0) \end{pmatrix}$

5. $\begin{pmatrix} (1,3) & (6,2) \\ (3,1) & (2,2) \end{pmatrix}$

6. $\begin{pmatrix} (8,4) & (1,5) \\ (3,3) & (10,2) \end{pmatrix}$

Example 4.22

(Evolutionarily Stable Strategies) Let us consider the following somewhat simplistic example from evolutionary biology. Consider a population consisting of an aggressive group (called Hawks) and a peaceful group (called Doves). When two Hawks have an encounter for some resource (food, mate, etc.), let us assign 10 points to the winner of the fight and −20 to the loser for the injuries suffered. (Hence, the expected outcome for each Hawk is $\frac{1}{2}(10) + \frac{1}{2}(-20) = -5$.) When a Hawk vies with a Dove for a resource, the Dove backs off without resistance. The Hawk receives 10 points and the Dove, 0. When Dove meets Dove, they eventually share the resource after some squabbling; let us assign 4 points to each Dove. (The assignment of points is reasonable, but somewhat arbitrary.) The following game matrix summarizes the situation.

	H	D
H	(−5,−5)	(10,0)
D	(0,10)	(4,4)

(See also Dawkins [1976] and Straffin [1993], Chapter 15.) If almost all the population consists of Doves, a Dove almost always gets 4 from an encounter, whereas a Hawk almost always gets 10. Hence, the Hawks will increase, and the given situation was not "evolutionarily stable". Likewise, if almost all the population consists of Hawks, then a Hawk almost always receives −5 from an encounter, whereas a Dove gets 0. In that case, the Doves increase and the given situation was not "evolutionarily stable". Hence, the question arises as to what original distribution of Hawks and Doves would give equal advantage to Hawks and Doves and thus lead to evolutionary stability. Let p be the original ratio of Hawks in the population; so, $1 - p$ is the original ratio of Doves. Then a Hawk should expect $-5p + 10(1 - p) = 10 - 15p$ points, and a Dove should expect $4(1 - p) = 4 - 4p$ points. For stability, the Hawk and Dove should have equal prospects, that is, $10 - 15p = 4 - 4p$, yielding the solution $p = 6/11$. Thus, 6/11 of the population should be Hawks and 5/11 Doves. We say that (6/11,5/11) is an *evolutionarily stable strategy* (ESS). Note that $X = (6/11,5/11)$, $Y = (6/11,5/11)$ is a mixed Nash equilibrium, which can be found by our standard technique. Each player gets a pay-off of 20/11 at this equilibrium, which is not as good as the pay-offs (4,4) for the pair of pure strategies (D,D). On the other hand, the pairs (H,D) and (D,H) are pure Nash equilibria with pay-offs (10,0) and (0,10). The joint randomized strategy that assigns $1/2$ to each of (H,D) and (D,H) would yield the still more advantageous pay-offs (5,5).

The significance of an ESS S is that, if the vast majority of the population plays that strategy against another strategy T, then the pay-off to an S player is at least as great as the pay-off to a T player. Hence, evolution should yield ESS's. A more nuanced and realistic exposition can be found in Weibull [1995].

Let us present a more general picture of a Hawk-Dove game. Let 2K measure the advantage obtained by a victory for a Hawk and let 2L measure the injury to a Hawk that loses such an encounter. Then the expected pay-offs for a Hawk vs. Hawk interaction are $(K - L,K - L)$. Let J be the advantage gained by a Dove in the stand-off that results from a Dove vs. Dove interaction. The total advantage for the two Doves is 2J and, since there is some wear and tear from the encounter, we may assume that $2J < 2K$. Then the game matrix would be:

	H	D
H	$(K - L, K - L)$	$(2K, 0)$
D	$(0, 2K)$	(J, J)

If, as in the example, p is the ratio of Hawks, the same kind of calculation as above yields $p = \frac{2K - J}{K + L - J}$, that is, the evolutionarily stable strategy is (X^*, Y^*), where $X^* = Y^* = (p, 1-p)$. The pay-off for each player is $\frac{(L - K)J}{K + L - J}$. One can check by the usual technique that this ESS (X^*, X^*) is a mixed Nash equilibrium.

Note that we can re-interpret our results in terms of the genetic composition within a species. Thus, instead of a population consisting of Hawks and Doves, we can think of a conflict within a species between aggressive (Hawk-like) and peaceful (Dove-like) tendencies. Then, since an ESS is a Nash equilibrium, an ESS (X^*, X^*) has the characteristic property that it can resist the incursion of a mutant subpopulation (X, X). An encounter between a member of the ESS population and a member of the mutant group will yield a pay-off for the ESS representative that is at least as great as that for the mutation.

Exercise 4.22

(a) Find evolutionarily stable strategies for the following Hawk vs. Dove games.

(i)

	H	D
H	(−1,−1)	(4,0)
D	(0,4)	(2,2)

(ii)

	H	D
H	(−9,−9)	(12,0)
D	(0,12)	(6,6)

(b) In the general Hawk–Dove game, examine the consequences of the three cases K > L, K = L, and K < L. In particular, identify any pure Nash equilibria, and determine when the game is biased in favor of Hawks or Doves.

4.5   The Nash arbitration procedure

We have seen how our attempts to find optimal strategies for the players in a two-person non-zero-sum game lead to a cooperative game. The set $\mathcal{R}^*$ of pay-offs corresponding to joint randomized strategies formed a convex polygon and its interior, and the upper northeastern boundary of $\mathcal{R}^*$ formed the negotiation set $\mathcal{N}$. That set consisted of the Pareto optimal points of $\mathcal{R}^*$, that is, those pay-offs (u,v) in $\mathcal{R}^*$ such that no other pay-off pair in $\mathcal{R}^*$ gave both players at least as much as their pay-offs in (u,v). (See the examples in Figures 4.3 to 4.5 and in Exercises 4.21 and 4.22.) In general, the negotiation set $\mathcal{N}$ will be a single point, a line segment of negative slope, or a polygonal line made up of such line segments. When $\mathcal{N}$ is a single point, that point gives us the optimal pay-offs and the pair of strategies corresponding to that point are the optimal strategies. But, in the other cases, we need a method for choosing one specific point out of the infinite set $\mathcal{N}$. There is no obvious way of doing this, since, as we move to the left (and upward) on $\mathcal{N}$, the pay-off to A decreases and the pay-off to B increases.

A method for choosing a point on $\mathcal{N}$ was offered in Nash [1950]. His approach, the Nash Arbitration Procedure, applies more generally to the problem of picking a point from a set consisting of a convex polygon and its interior. In our applications, that set will be $\mathcal{R}^*$. Nash proceeds by listing conditions that his procedure must satisfy, giving a precise definition of his procedure, and showing that his procedure is the only one that fulfill his conditions.

A peculiar and somewhat arbitrary facet of Nash's procedure is that it depends on an initial choice of a point in $\mathcal{R}^*$ that serves as a kind of default position. It will be referred to as the *status quo* and will be denoted *SQ*. It can be thought of as the pay-offs that the players will receive if they reject the point determined by Nash's procedure. We shall discuss the selection of *SQ* later. (Likely candidates are the *security levels*, the values that each player can guarantee for himself by playing a suitable strategy that depends in the usual way on his pay-offs alone.[14]) SQ will sometimes be written as the pair (u_0, v_0). We shall use Φ as a variable for an arbitration procedure, so that $\Phi [\mathcal{R}^*, (u_0, v_0)]$ will denote the point in $\mathcal{R}^*$ that Φ selects when it is applied to $\mathcal{R}^*$ and the SQ (u_0, v_0).

We shall abbreviate $\Phi [\mathcal{R}^*, (u_0, v_0)]$ by (u^*, v^*). As before, we shall let $\mathcal{N}$ denote the set of Pareto optimal points of $\mathcal{R}^*$. Now let us list the conditions that Nash requires of an arbitration procedure Φ.

1. (u_0, v_0) is in $\mathcal{R}^*$, its image (u^*, v^*) is in $\mathcal{N}$, and $u_0 \le u^*$, $v_0 \le v^*$.
2. **(Linearity)** If $\mathcal{R}^{**}$ is a polygonal region obtained from $\mathcal{R}^*$ by a positive linear transformation f, then the result of applying Φ to $\mathcal{R}^{**}$ with respect to the new SQ $f(u_0, v_0)$ is $f(u^*, v^*)$. More precisely, if $f(u,v) = (au + b, cv + d)$ with $a > 0$ and $c > 0$, then $\Phi [\mathcal{R}^{**}, (au_0 + b, cv_0 + d)] = (au^* + b, cv^* + d)$.
3. **(Symmetry)** Assume that the region $\mathcal{R}^*$ is symmetric with respect to the diagonal $u = v$[15] and that (u_0, v_0) is on the diagonal, that is, $u_0 = v_0$. Then (u^*, v^*) will be on the diagonal, that is, $u^* = v^*$.
4. **(Independence of Irrelevant Alternatives)** If $\mathcal{R}^{**}$ is a convex polygonal region included within $\mathcal{R}^*$ and both (u_0, v_0) and (u^*, v^*) are in $\mathcal{R}^{**}$, then $\Phi [\mathcal{R}^{**}, (u_0, v_0)] = (u^*, v^*)$.

The Linearity condition ensures that the arbitration procedure is not affected by a change of units, and the Symmetry condition seems plausible. The Independence of Irrelevant Alternatives tells us that, if (u^*, v^*) is the result of an application of the arbitration procedure and we apply the procedure to a smaller polygonal region containing (u^*, v^*) and the original SQ, then the procedure still yields (u^*, v^*) if we keep the original SQ. Of all the conditions, this is the one that is the least convincing.

The Nash Arbitration Procedure Nash gives the following explicit arbitration procedure Φ_N. If the status quo SQ is (u_0, v_0), then $\Phi_N[\mathcal{R}^*, (u_0, v_0)]$ is defined to be the point (u, v) in the negotiation set N that maximizes the product $(u - u_0)(v - v_0)$.

Nash proves that his arbitration procedure satisfies conditions (1) to (4) and that it is the only procedure that does so. See Exercise 4.29.

Let us see how the Nash arbitration procedure works in a few cases.

---

[14]For example, if the game matrix is

$$\begin{pmatrix} (2,5) & (3,4) \\ (4,1) & (1,2) \end{pmatrix}$$

then A and B have matrices

$$\begin{pmatrix} 2 & 3 \\ 4 & 1 \end{pmatrix} \quad \text{and} \quad \begin{pmatrix} 5 & 1 \\ 4 & 2 \end{pmatrix}$$

with values 5/2 and 3, respectively.

[15]That is, if (u,v) is in $\mathcal{R}^*$, then (v,u) also is in $\mathcal{R}^*$.

Example 4.23

In Example 4.19, the negotiation set $\mathcal{N}$ is the line segment RS connecting $R(1,2)$ to $S(2,1)$. See Figure 4.13. Let us choose $SQ = (u_0, v_0) = (1,1)$, which assigns to A and B the worst pay-offs they each could get on RS. An equation for RS is $v = -u + 3$, for $1 \leq u \leq 2$. Hence, the product

$$g(u,v) = (u - u_0)(v - v_0) = (u - 1)(v - 1) = (u - 1)(-u + 2)$$

$$= -u^2 + 3u - 2$$

Then, by elementary calculus, the maximum occurs when the derivative is 0, that is, when $u = 3/2$; the corresponding value of v is 3/2. So, the point (u^*, v^*) determined by Φ_N is $(3/2, 3/2)$, which, in this case, is the midpoint of RS.

Exercise 4.23

If the negotiation set is a line RS and the maximum of $(u - u_0)(v - v_0)$ occurs at a point Q of the line RS that is outside of the segment RS, show that the maximum value of $(u - u_0)(v - v_0)$ on the segment RS occurs at the endpoint of RS closest to Q (and, therefore, that endpoint is the result of the Nash arbitration procedure).

In Example 4.23, we could have used the following shortcut for finding the point (u^*, v^*) determined by Nash's arbitration procedure.

Shortcut. If the negotiation set $\mathcal{N}$ is a line segment RS with slope $-m$, then (u^*, v^*) is the intersection of RS with the line with slope m passing through $SQ = (u_0, v_0)$. (However, if the result is outside segment RS, then (u^*, v^*) is the endpoint of RS closer to the intersection.)

In Example 4.23, RS has slope $-m = -1$. The line with slope $m = 1$ passing through $(1,1)$ has the equation $v = u$. The line RS has the equation $v = -u + 3$. So, (u^*, v^*) is obtained by solving the pair of equations $v = u$ and $v = -u + 3$. This yields $u^* = 3/2 = v^*$.

Exercise 4.24

Prove the Shortcut. (Hint: The equation for RS has the form $v = -mu + b$, and the equation for the line through SQ is $v = mu + v_0 - mu_0$. So, the intersection (u_1, v_1) satisfies $2mu_1 = b - v_0 + mu_0$ and $2v_1 = b + v_0 - mu_0$. The function to be maximized is

$$(u - u_0)(v - v_0) = (u - u_0)(-mu + b - v_0)$$

and the maximum occurs at

$$u^* = (mu_0 + b - v_0)/2m. \text{ Hence, } u^* = u_1.$$

Interlude (Security levels) In choosing the status quo point SQ, it would seem to be plausible for A and B to get values that are at least as big as the values that they can be sure of obtaining, the so-called *security levels*. Define these values v_A and v_B as follows:

$$v_A = \max_X \min_Y P_A(X,Y)$$

where X varies over all mixed strategies for A, and Y varies over all mixed strategies for B, and, similarly, $v_B = \max_X \min_Y P_B(X,Y)$. Instead of *security levels*, the numbers v_A and v_B are sometimes called the *maximin values* for A and B. (Analogous definitions can be given when the number of players is greater than two.)

The security levels can be obtained by the following procedure. For player A, look at the matrix of pay-offs for A alone. Then the security level v_A is the value of this matrix, considered as the matrix of a two-person zero-sum game; in fact, this value is the maximum that A can expect in the worst possible case, that is, the case when B tries to minimize A's pay-off. Similarly, the security level v_B can be obtained by applying the same procedure to B's pay-off matrix.

Example 4.24

1. For the matrix

$$\begin{pmatrix} (4,3) & (2,5) \\ (0,4) & (3,3) \end{pmatrix}$$

first look at A's pay-off matrix

$$\begin{pmatrix} 4 & 2 \\ 0 & 3 \end{pmatrix}$$

By Theorem 2.14, its value is 12/5 (attained at $X = (3/5,2/5)$). Thus, $v_A = 12/5$. B's pay-off matrix is

$$\begin{pmatrix} 3 & 4 \\ 5 & 3 \end{pmatrix}$$

(Note that we have to transpose rows and columns.) Its value is $(-11)/(-3) = 11/3$ (attained at $Y = (2/3,1/3)$). So, $v_B = 11/3$.

2. Consider the matrix

$$\begin{pmatrix} (-1,2) & (3,4) \\ (1,0) & (4,1) \end{pmatrix}$$

A's pay-off matrix is

$$\begin{pmatrix} -1 & 3 \\ 1 & 4 \end{pmatrix}$$

This has a saddle point 1, which is the value v_A. B's pay-off matrix is

$$\begin{pmatrix} 2 & 0 \\ 4 & 1 \end{pmatrix}$$

This also has a saddle point 1, which is, therefore, v_B.

Exercise 4.25
Find the security levels for the matrices in Examples 4.20 and 4.21, and Exercises 4.21 (4, 5, 6).

Example 4.25
Let us consider the game given by the matrix

$$\begin{pmatrix} (3,1) & (2,2) \\ (0,5) & (3,4) \end{pmatrix}$$

of Example 4.20. The negotiation set is the line segment RS connecting $R = (0,5)$ and $S = (3,4)$. (See Figure 4.4.) Let us take SQ to be the point (9/4,3) given by the security levels for A and B. 9/4 is the security level for A in his matrix

$$\begin{pmatrix} 3 & 2 \\ 0 & 3 \end{pmatrix}$$

and 3 for B in his matrix.

$$\begin{pmatrix} 1 & 5 \\ 2 & 4 \end{pmatrix}$$

Let us use the Shortcut. Segment RS is given by the equation $v = -\frac{1}{3}u + 5$ for $0 \le u \le 5$. The line through $SQ = (9/4,3)$ with slope $\frac{1}{3}$ has the equation $v = \frac{1}{3}u + \frac{9}{4}$. The intersection occurs at (33/8,29/8), which is inside the segment RS, and it is, therefore, the point chosen by the arbitration procedure.

Example 4.26
Let us consider the game given by the matrix

$$\begin{pmatrix} (4,3) & (2,5) \\ (1,1) & (5,0) \end{pmatrix}$$

of Example 4.21(a). The negotiation set is the broken line RST connecting $R = (2,5)$, $S = (4,3)$, $T = (5,0)$ (See Figure 4.5.) Let us take two different values of SQ.

1. Let $SQ = (3,5/3)$, the security levels. RS has the equation $v = -u + 7$ for $2 \leq u \leq 4$. Then $(u - 3)(v - 5/3) = (u - 3)(-u + 16/3) = -u^2 + 25/3\, u - 16$, so that the maximum occurs at $u = 25/6$, which gives a point outside segment RS. Hence, the maximum on RS occurs at the nearest endpoint S. ST has the equation $v = -3u + 15$ for $4 \leq u \leq 5$. Then $(u - 3)(v - 5/3) = (u - 3)(-3u + 40/3) = -3u^2 + 67/3\, u - 40$. So, the maximum occurs at $u = 67/18$, giving us a point outside of segment ST. Thus, the maximum on ST occurs at the nearest endpoint S. Since the maximum occurs at S for both RS and ST, the arbitration procedure selects $S = (4,3)$.
2. Let SQ be the vertex $(1,1)$. For RS, let us use the Shortcut. We get the equations $v = -u + 7$ and $v = u$. The common solution is $(7/2,7/2)$, which is inside segment RS. For ST, we must solve the equations $v = -3u + 15$ and $v = 3u - 2$. The solution yields $u = 17/6$, so that the maximum occurs outside segment ST. So, the maximum on ST occurs at the nearest endpoint S. But, since the maximum on RS occurs inside the segment, that maximum is greater than the value at S. Hence, the maximum occurs at $(7/2,7/2)$, which is, therefore, the value (u^*,v^*) assigned by the arbitration procedure. Note that this differs from the result in (1), so that the result of the arbitration procedure may depend on the choice of SQ.

Exercise 4.26
Apply Nash's arbitration procedure to the following examples:

1. Example 4.1 with (i) $SQ = (0,2)$, (ii) $SQ = (3,1)$, the security levels
2. Example 4.2 with arbitrary SQ
3. Example 4.3 with $SQ = (5/2,18/5)$, the security levels
4. Example 4.6 with SQ determined by the security levels
5. Example 4.8 with SQ determined by the security levels

Exercise 4.27
Apply Nash's arbitration procedure in the following cases:

1. The matrix

$$\begin{pmatrix} (1,3) & (0,-1) \\ (2,0) & (-1,3) \end{pmatrix}$$

with $SQ = (1/2,9/7)$.
2. The matrix

$$\begin{pmatrix} (0,1) & (1,6) \\ (3,5) & (5,2) \end{pmatrix}$$

with (i) $SQ = (0,1)$, (ii) $SQ = (1,2)$.

3. The matrix

$$\begin{pmatrix} (2,8) & (5,0) \\ (3,6) & (4,3) \end{pmatrix}$$

with $SQ = (2,0)$.

Exercise 4.28
Consider the game matrix

$$\begin{pmatrix} (1,3) & (-1,-3) \\ (-3,-1) & (3,1) \end{pmatrix}$$

1. Find the pure and mixed Nash equilibria.
2. Show that the pay-offs from the mixed Nash equilibrium are $(0,0)$.
3. Show that the security levels are $(0,0)$.
4. Show that the Nash arbitration procedure yields the point $(2,2)$ when $SQ = (0,0)$.
5. Show that B has a threat possibility that should force the result $(1,3)$, which is much better than the result for B in part (4). Does this cast doubt on the reasonableness of the Nash arbitration procedure? The Nash arbitration procedure with SQ given by the security levels is called the *Shapley procedure* in Luce-Raiffa [1957], Section 6.8.
6. Apply the Shapley procedure to the games defined by the matrices in the following cases: (i) Exercise 4.21(4) (ii) Exercise 4.21 (5) (iii) The Prisoner Dilemma Game

$$\begin{pmatrix} (0,0) & (5,-5) \\ (-5,5) & (2,2) \end{pmatrix}$$

(iv) The Game of Chicken

$$\begin{pmatrix} (-50,-50) & (4,-2) \\ (-2,4) & (-2,-2) \end{pmatrix}$$

(v) The Battle of the Sexes matrix

$$\begin{pmatrix} (6,3) & (4,4) \\ (0,0) & (3,6) \end{pmatrix}$$

(vi) The original Battle of the Sexes matrix

$$\begin{pmatrix} (2,1) & (-1,-1) \\ (-1,-1) & (1,2) \end{pmatrix}$$

Exercise 4.29

1. Verify that the Nash arbitration procedure Φ_N satisfies Nash's conditions (1) to (4).
2. Fill in the details of the following sketch of a proof that any arbitration procedure Ψ satisfying conditions (1) to (4) is identical with Nash's procedure Φ_N.[16] We are given as usual a polygonal region $\mathcal{R}^*$ and status quo $SQ = (u_0,v_0)$, with $\mathcal{R}^*$ bounded on the northeast by the negotiation set consisting of a polygonal line. Carry out a positive linear mapping T_1 that moves the SQ (u_0,v_0) to the origin $(0,0)$, and transforms $\mathcal{R}^*$ into a similar region $\mathcal{R}_1$. Let (u_1,v_1) be the result of applying Nash's procedure to $\mathcal{R}_1$ and $(0,0)$, that is, $(u_1,v_1) = \Phi_N[\mathcal{R}_1, (0,0)]$. So, (u_1,v_1) is a point on the polygonal line forming the northeastern boundary of $\mathcal{R}_1$. Now apply a positive linear mapping T_2 that keeps $(0,0)$ fixed and moves (u_1,v_1) to $(1,1)$. Let $\mathcal{R}_2$ be the polygonal region that is the image of $\mathcal{R}_1$ under T_2. Now it suffices to show: (*) $(1,1) = \Psi[\mathcal{R}_2, (0,0)]$. For, if we assume (*), then we can apply the positive linear mapping T_2^{-1} that is, the inverse of T_2. By the linearity condition (2) for Ψ with respect to T_2^{-1}, we get $(u_1,v_1) = \Psi[\mathcal{R}_1, (0,0)]$. Again, by the linearity condition for Ψ with respect to the positive linear mapping T_1^{-1}, we obtain

$$(u_0 + u_1, v_0 + v_1) = \Psi[\mathcal{R}^*, (u_0,v_0)].$$

But, by the linearity condition for Φ_N

$$(u_0 + u_1, v_0 + v_1) = \Phi_N[\mathcal{R}, (u_0,v_0)].$$

Hence, Φ_N is identical with Ψ. To prove (*), use the following reasoning. Let $\mathcal{R}_3$ be the *symmetrization* of $\mathcal{R}_2$, that is, the symmetric set[17] obtained by adding to $\mathcal{R}_2$ its reflection in the diagonal line $u = v$. By the symmetry condition (3) for Ψ, $(1,1) = [\mathcal{R}_3, (0,0)]$, since $(1,1)$ is the only Pareto optimal point on the diagonal line. Hence, by condition (4) for the polygonal regions $\mathcal{R}_2$ and $\mathcal{R}_3$, $(1,1) = \Psi[\mathcal{R}_2, (0,0)]$.

The pay-offs (u^*,v^*) that result from the application of Nash's arbitration procedure depend on the initial choice of a status quo point SQ. To be fair

---

[16]More detailed and rigorous proofs may be found in Owen [1968], Section VII.2, or Morris [1994], pp. 134–142.
[17]Recall that a planar set S is symmetric if, whenever (u,v) is in S, (v,u) also is in S.

to both players, SQ should reflect the relative strengths of the players, including the threat possibilities. However, there is no general agreement on how
to measure those strengths.

Nash [1953] attacked this problem in the following way. For a given
game G, Nash showed how to find strategies X^* and Y^* and numbers u^* and
v^* such that:

1. If A plays X^*, he is guaranteed a pay-off of at least u^*, and if B plays
    Y^*, he is guaranteed a pay-off of at least v^*.
2. X^* and Y^* form an equilibrium pair and, for any other equilibrium
    pair X^{**} and Y^{**}, A will have a pay-off of at least u^* and B will have
    a pay-off of at least v^*.

(u^*,v^*) can be proved to be Pareto optimal and in the negotiation set. Nash
then defines the following non-cooperative game G^*. As their first moves,
A and B simultaneously choose mixed strategies X and Y of G. Knowing
both of those moves, players A and B simultaneously choose numbers d_A
and d_B. The pay-offs are now determined in the following manner:

1. If (d_A,d_B) is in the region $\mathcal{R}^*$ of pay-offs determined by joint random-
    ized strategies, then A gets d_A and B gets d_B;
2. If (d_A,d_B) is not in $\mathcal{R}^*$, then A receives $P_A(X,Y)$ and B receives $P_B(X,Y)$.

Nash shows that (X^*,u^*) and (Y^*,v^*) form an equilibrium pair for this game
and that the corresponding pair of pay-offs is Pareto optimal.[18] He also gives
arguments that this is, in a certain sense, an optimal solution, but these
arguments are not universally accepted.[19]

4.6   Games with two or more players

Such games can be given either in extensive or in normal form. The extensive
form describes the possible sequences of moves that may occur during the play
of the game, whereas the normal form only lists the strategies of the various
players and the pay-offs that result for each choice of strategies. The extensive
form may provide more information and a better insight into the complexities of a game, but the normal form is generally simpler and easier to study
and it is the version that we shall deal with.

When there are more than two players, there is no visually satisfying way
of showing the normal form. One can specify a method for listing all choices
of strategies, one for each player, and the corresponding pay-offs, but such a
list usually offers no helpful geometric picture of the properties of the game.

---

[18]Some idea of the reasoning may be found in Owen [1968], Section VII.3.
[19]See the discussion in Luce-Raiffa [1957], Section 6.9, where Nash's scheme is called Nash's
Extended Bargaining Model. Consult Jones [1980] for a more precise treatment with proofs.

A partial exception is the case where there are three players and one of them, say Player A, has a small number of strategies, say $A_1,..., A_r$. Let the other two players B and C have strategies $B_1,..., B_m$ and $C_1,..., C_n$, respectively. For each A_i, we can draw an $m \times n$ matrix, with the m rows corresponding to $B_1,..., B_m$ and the n columns corresponding to $C_1,..., C_n$. The entry for the jth row and kth column will be the triple of numbers that are the pay-offs to A, B, and C, respectively, when the players play the strategies A_i, B_j, C_k.

Example 4.27

Assume that each of A, B, and C has two strategies. Then the following pair of 2×2 matrices, one for strategy A_1 and one for strategy A_2, determines a game in normal form.

A_1	C_1	C_2	A_2	C_1	C_2
B_1	(1,0, −1)	(2,1,0)	B_1	(4,1,2)	(−2,0,1)
B_2	(1,1,1)	(2,3,0)	B_2	(2,2,3)	(0,1,1)

For example, when A, B, and C play strategies A_2, B_1, and C_2, respectively, their pay-offs are −2, 0, and 1, respectively. To look for pure Nash equilibria, we can use a generalization of the technique employed for games with two players. Attach # as a superscript to certain entries according to the following rules.

For A's pay-offs, for each (B_j, C_k) look at the *first* components of the two triples corresponding to (A_1, B_j, C_k) and (A_2, B_j, C_k), and attach a # to the larger of the two (or to both if they are equal). For example, the triples corresponding to (A_1, B_1, C_1) and (A_2, B_1, C_1) are (1,0, −1) and (4,1,2) ; so, we would attach # to the component 4. For B's pay-offs, look at the triples in each of the four *columns* and attach # to the larger of the *second* components in each column. For example, if we look at the second components in the two triples (2,1,0) and (2,3,0) under C_2 in the A_1 matrix, we would attach # to the component 3. Similarly, for C's pay-offs, look at the triples in each of the four *rows* and attach # to the larger of the *third* components in each row. For example, if we look at the third components in the two triples (4,1,2) and (−2,0,1) in the B_1-row of the A_2 matrix, we would attach # to the component 2. A Nash equilibrium occurs at any triple in which all three components have # attached. Our example now has the following appearance:

A_1	C_1	C_2	A_2	C_1	C_2
B_1	(1,0, −1)	$(2^\#,1,0^\#)$	B_1	$(4^\#,1,2^\#)$	(−2,0,1)
B_2	$(1,1^\#,1^\#)$	$(2^\#,3^\#,0)$	B_2	$(2^\#,2^\#,3^\#)$	$(0,1^\#,1^\#)$

Thus, there is a unique pure Nash equilibrium (2,2,3) at (A_2, B_2, C_1). (In this particular example, we could have arrived at the same conclusion via dominance relations.)

Example 4.28

Let us change the game in Example 4.27 by changing the (A_2,B_2,C_1) entry from (2,2,3) to (2,2,0).

A_1	C_1	C_2	A_2	C_1	C_2
B_1	(1,0, −1)	(2,1,0)	B_1	(4,1,2)	(−2,0,1)
B_2	(1,1,1)	(2,3,0)	B_2	(2,2,0)	(0,1,1)

It is easy to see that there is no longer any pure Nash equiibrium. However, Nash's Theorem assures us that there must be at least one mixed Nash equilibrium, and we can try to find it by using our earlier method that employed partial derivatives. Let $X = (x,1 − x)$, $Y = (y,1 − y)$, and $Z = (z,1 − z)$ be mixed strategies for A, B, and C. Then the pay-offs $P_A(x,y,z)$, $P_B(x,y,z)$, and $P_C(x,y,z)$ can be evaluated as follows:

$$P_A(x,y,z) = x[yz + 2y(1 − z) + (1 − y)z + 2(1 − y)(1 − z)]$$

$$+ (1 − x)[4yz − 2y(1 − z) + 2(1 − y)z]$$

$$= −4xyz + 2xy − 3xz + 4yz + 2x − 2y + 2z$$

Similarly, we get $P_B(x,y,z) = xyz − xy − 3xz + 2x − y + z + 1$ and

$$P_C(x,y,z) = −4xyz + 2xz + 2yz − x − z + 1$$

Now, $\partial P_A/\partial x = −4yz + 2y − 3z + 2$. Setting this equal to 0 and solving for z, we get $z = 2(y + 1)/(4y + 3)$. (Using $0 \le y \le 1$, we deduce that $0 < z < 1$.) From the formula for P_B, we obtain $\partial P_B/\partial y = xz − x − 1$. Since this is always negative, P_B is decreasing on the interval [0,1] and the maximum of P_B should occur at $y = 0$. Since $z = 2(y + 1)/(4y + 3)$, we get $z = 2/3$. In addition, the formula for P_C yields $\partial P_C/\partial z = −4xy + 2x + 2y − 1$. At a maximum for P_C, $\partial P_C/\partial z = 0$, that is, $−4xy + 2x + 2y − 1 = 0$. From $y = 0$, $z = 2/3$ it follows that $x = 1/2$. Thus, the mixed Nash equilibrium is formed by $X = (1/2,1/2)$, $Y = (0,1)$, $Z = (2/3,1/3)$ and the corresponding pay-offs are (4/3,5/3,1/2). From the formulas for P_A, P_B, and P_C, it is easily verified that we actually have found a Nash equilibrium.

Exercise 4.30

For the following three-person games, find Nash equilibria.

(a)

A_1	C_1	C_2	A_2	C_1	C_2
B_1	(1,0,2)	(2, −1,0)	B_1	(2,1,3)	(4,1,2)
B_2	(0,4,3)	(3,1,2)	B_2	(2,2,2)	(0,0,1)

(b)

A_1	C_1	C_2	A_2	C_1	C_2
B_1	(5,1,2)	(0,0,3)	B_1	(4,2,3)	(1,−1,1)
B_2	(1,2,4)	(1,2,1)	B_2	(2,0,4)	(3,1,2)

(c)

A_1	C_1	C_2	A_2	C_1	C_2
B_1	(1,0,1)	(−1,1,0)	B_1	(0,2,1)	(0,0,1)
B_2	(0,1,1)	(1,0,0)	B_2	(−1,1,1)	(0,0,1)

(d) In this problem, player A has three strategies.

A_1	C_1	C_2	A_2	C_1	C_2	A_3	C_1	C_2
B_1	(0,1,0)	(1,0,0)	B_1	(2,0,1)	(0,0,1)	B_1	(1,1,1)	(0,1,1)
B_2	(−1,1,0)	(0,0,0)	B_2	(0,1,0)	(0,1,1)	B_2	(1,0,1)	(0,0,−1)

Use $(x_1, x_2, 1 - x_1 - x_2)$ to designate an arbitrary strategy for player A.

(e) (Three-person Prisoners' Dilemma)

A_1	C_1	C_2	A_2	C_1	C_2
B_1	(3,3,3)	(2,2,5)	B_1	(5,2,2)	(4,0,4)
B_2	(2,5,2)	(0,4,4)	B_2	(4,4,0)	(1,1,1)

(f) Player A again has three strategies.

A_1	C_1	C_2	A_2	C_1	C_2	A_3	C_1	C_2
B_1	(2,0,1)	(1,0,2)	B_1	(1,1,1)	(0,2,0)	B_1	(−1,0,0)	(0,0,1)
B_2	(0,1,0)	(1,1,0)	B_2	(0,1,1)	(1,0,1)	B_2	(0,2,0)	(1,3,0)

Example 4.29
Consider the following three-person zero-sum game.

A_1	C_1	C_2	A_2	C_1	C_2
B_1	(1,0,−1)	(0,1, −1)	B_1	(−1,0,1)	(1,0, −1)
B_2	(0,0,0)	(−1,1,0)	B_2	(0,−1,1)	(0,1, −1)

There are two pure Nash equilibria (1,0, −1) at (A_1, B_1, C_1) and (0,0,0) at (A_1, B_2, C_1). The usual computation with partial derivatives yields a mixed Nash equilibrium $X = (1,0)$, $Y = (1/2, 1/2)$, and $Z = (1/2, 1/2)$, with pay-offs (0,1/2, −1/2). Note that the fact that the game is zero-sum does not introduce the simplifications that were present in two-person zero-sum games. (For example, the pay-offs at different Nash equilibria need not be identical.)

4.7 Coalitions

In games with three or more players, it is possible for two or more of the players to form a *coalition* against the other player or players. The players in a coalition can coordinate their choice of strategies to obtain an advantage against their opponents.

Let us look, for example, at the game in Example 4.29. Consider first what happens when B and C form a coalition against A. We obtain the following pay-off table for A.

	B_1C_1	B_1C_2	B_2C_1	B_2C_2
A_1	1	0	0	−1
A_2	−1	1	0	0

Since the original game was zero-sum, it is only necessary to give A's pay-offs. Now we can apply the known techniques for two-person zero-sum games. By dominance, we can drop the second and third columns, since they dominate the fourth column. So we get:

	B_1C_1	B_2C_2
A_1	1	−1
A_2	−1	0

There is no saddle point. Hence, by Theorem 2.14, the equilibrium mixed strategy consists of (1/3,2/3) for A and (1/3,0,0,2/3) for BC. The value of the game is −1/3, so that A can be sure of losing at most 1/3, and the BC coalition can be sure of gaining at least 1/3. We shall indicate this by writing $v(A) = -1/3$ and $v(BC) = 1/3$. When A and BC play their optimal mixed strategies, then the pay-offs will be:

$$(1/3)(1/3)(1,0, -1) + (1/3)(2/3)(-1,1,0) + (2/3)(1/3)(-1,0,1)$$

$$+ (2/3)(2/3)(0,1, -1) = (-1/3,2/3, -1/3)$$

Thus, A would lose 1/3, B would win 2/3 and C would lose 1/3.

What happens when A and C form a coalition against B? We get the following matrix of pay-offs for B.

	A_1C_1	A_1C_2	A_2C_1	A_2C_2
B_1	0	1	0	0
B_2	0	1	− 1	1

Dominance eliminates all columns but the third, and then dominance leaves (B_1, A_2C_1). So, the value is 0. Thus, $v(B) = 0$ and $v(AC) = 0$. The pay-offs at

(A_2, B_1, C_1) are $(-1,0,1)$. Thus, A loses 1, B neither wins nor loses, and C wins 1. Now consider a coalition of A and B against C. The table of pay-offs for C is:

	A_1B_1	A_1B_2	A_2B_1	A_2B_2
C_1	−1	0	1	1
C_2	−1	0	−1	−1

Dominance eliminates the second row, and then dominance leads to (C_1, A_1B_1). Thus, the value is −1, $v(C) = -1$ and $v(AB) = 1$. The pay-offs are $(1,0,-1)$. So, A wins 1, B breaks even, and C loses 1.

Which coalitions are likely to form? Now, A can expect to lose 1 when he joins with C, to win 1 when he joins with B, and to lose 1/3 when he plays alone. Hence, A prefers a coalition with B. As for B, he wins 2/3 when he joins with C, breaks even when he joins A, and breaks even when he plays alone. So, B prefers a coalition with C. Finally, C loses 1/3 when he plays with B, wins 1 when he joins with A, and loses 1 when he plays alone. Thus, C prefers to play with A. Therefore, A wants to unite with B, B wants to unite with C, and C wants to unite with A. Apparently then no coalition will form. However, notice that, when playing alone, A loses 1/3 and B breaks even, but their coalition wins 1 when playing against C. Therefore, A and B would be better off if they formed a coalition and then divided the pay-off of 1 evenly; they would each get $^1/_2$, and C would lose 1. Such an arrangement is possible only if binding side payments are allowed. But it is not clear that A and B would be happy with the even division of their pay-off. Obviously this deal-making might never reach a conclusion.

Exercise 4.31
For the following three-person zero-sum game, find the Nash equilibria and the corresponding pay-offs. Find the pay-offs for the various coalitions and determine which coalitions (if any) are likely to be formed.

A_1	C_1	C_2	A_2	C_1	C_2
B_1	(2,−1,−1)	(−1,0,1)	B_1	(−2,1,1)	(0,2,−2)
B_2	(0,0,0)	(0,1,−1)	B_2	(1,−1,0)	(1,0,−1)

Example 4.30
Consider the following three-person, non-zero-sum game.

A_1	C_1	C_2	A_2	C_1	C_2
B_1	(1,2,−1)	(−1,1,0)	B_1	(0,3,1)	(1,−1,2)
B_2	(0,1,0)	(2,0,1)	B_2	(2,1,0)	(0,0,1)

There are no pure Nash equilibria. The reader should carry out the technique involving partial derivatives and verify that $X = (1/2,1/2)$, $Y = (1/2,1/2)$, $Z = (0,1)$ form a mixed Nash equilibrium, with pay-offs $(1/2,0,1)$. (Note that strategy C_2 dominates C_1.) Let us see how the BC coalition fares. The relevant matrix is:

$$
\begin{array}{ccc}
 & B_1C_2 & B_2C_2 \\
A_1 & (-1,1) & (2,1) \\
A_2 & (1,1) & (0,1)
\end{array}
$$

Here, the second component of each entry is the sum of the B and C pay-offs. If we look at the matrix for A's pay-offs, the value $v(A)$ is $^1/_2$ and its optimal mixed strategy is (1/4,3/4). The BC matrix has all entries equal to 1; so, $v(BC) = 1$. The final pay-offs are $(1/4)(1/2)(-1,1,0) + (1/4)(1/2)(2,0,1) + (3/4)(1/2)(1,-1,2) + (3/4)(1/2)(0,0,1) = (1/2, -^1/_4, 5/4)$. For B versus coalition AC, we get the matrix

$$
\begin{array}{ccc}
 & A_1C_2 & A_2C_2 \\
B_1 & (1,-1) & (-1,3) \\
B_2 & (0,3) & (0,1)
\end{array}
$$

where the second component of each entry is the sum of the A and C pay-offs. Looking at the matrix of B pay-offs, we have a saddle point with value 0, so that B plays strategy B_2. For the matrix of the $A + C$ sums, we get a value of 5/3 and mixed strategy (1/3,2/3). So, $v(B) = 0$ and $v(AC) = 5/3$. The final pay-offs are: $(1/3)(2,0,1) + (2/3)(0,0,1) = (2/3,0,1)$. A computation for C versus AB yields $v(C) = 1$, $v(AB) = 2$, and final pay-offs (2,0,1).

Now let us see which coalitions are preferred. A obtains pay-offs of 2 with B, 2/3 with C, and $^1/_2$ alone. B obtains pay-offs of 0 with A, $-^1/_4$ with C, and 0 alone. C obtains pay-offs of 5/4 with B, 1 with A, and 1 alone. Hence, A prefers a union with B, B prefers a union with A or no one, and C prefers a union with B. So, A and C might bid for B's cooperation. That cooperation is more important to A than to C, and A would have greater resources available to offer as a side payment to B. Thus, it seems certain that A and B will form a coalition, but a final resolution of the negotiations between A and B is not clear.

Exercise 4.32
For the following three-person games, find the Nash equilibria and the corresponding pay-offs, and investigate what is likely to happen when co-alitions are allowed.

(a)

$$
\begin{array}{cccc}
A_1 & C_1 & C_2 \\
B_1 & (-1,0,0) & (0,1,0) \\
B_2 & (0,0,1) & (1,0,0)
\end{array}
\qquad
\begin{array}{cccc}
A_2 & C_1 & C_2 \\
B_1 & (1,0,1) & (0,0,0) \\
B_2 & (0,0,-1) & (0,1,0)
\end{array}
$$

(b)

$$
\begin{array}{cccc}
A_1 & C_1 & C_2 \\
B_1 & (2,-1,0) & (0,1,0) \\
B_2 & (0,1,0) & (2,0,-1)
\end{array}
\qquad
\begin{array}{cccc}
A_2 & C_1 & C_2 \\
B_1 & (0,3,-2) & (0,-1,2) \\
B_2 & (0,0,1) & (1,0,0)
\end{array}
$$

(c) Here, A has three strategies.

A_1	C_1	C_2	A_2	C_1	C_2	A_3	C_1	C_2
B_1	(1,2, −1)	(0,3,1)	B_1	(3,1,2)	(2,2,2)	B_1	(0, −2,1)	(−2,1,2)
B_2	(2,3,0)	(1,−1,2)	B_2	(0,0, −1)	(4,0,−2)	B_2	(2,0,0)	(1,1,1)

4.8   Games in coalition form

In our initial study of games, we considered games in extensive form. The description of such games specified the moves that the players can make and the rules for determining the pay-offs when a game ends. It turned out to be convenient to abstract from such a detailed description by listing the strategies available to the players and constructing the corresponding matrix of pay-offs. This simplified form of the game was referred to as the game in normal form. Now, in our study of games with more than two players, we have been lead to the notion of coalitions of players and the pay-offs the players in such coalitions can realize for their coalition as a whole by suitable choices of strategies. There may be binding agreements that will govern the division of the pay-offs among the players in a coalition, but such agreements are not specified by the rules of the game.

Definition

A *game in coalition form* involves a finite set of players $N = \{p_1,..., p_k\}$. Any subset of the players is called a *coalition*. There is a function v that assigns a real number $v(S)$ to each coalition S. The only restriction on v is that $v(\emptyset) = 0$, where $\emptyset$ denotes the empty set. The number $v(S)$ may be called the *value* of the game for S. As notation for the game, we will sometimes use the pair (N,v).

Example 4.31

(*Characteristic Function*) Consider an ordinary matrix game among a set N of k players. Each player has a finite number of strategies and, for each choice of strategies $S_1,..., S_k$ by the players, there are pay-offs $P_j(S_1,..., S_k)$ for the players $p_j, j = 1,..., k$. For each non-empty subset S of N, let $v(S)$ be the support level of S. Recall how this is defined. Consider the following two-person game between two players A and B. A is the set S, and B is the complementary set $N - S$. A strategy for player A is a choice of strategies for each of the players in S, and a strategy for B is a choice of strategies for each of the players in $N - S$. For each strategy for A (which is a list L_1 of strategies for the players in S) and each strategy for B (which is a list L_2 of strategies for the players in $N - S$), the entry in our new game matrix is obtained by finding the pay-offs determined by L_1 and L_2 and taking the sum of just those pay-offs for the players in S. Define $v(S)$ to be the value of this matrix game for player A. (We are treating this game as a zero-sum game in order to get the maximum A can expect against all possible opposition.) The function v is called the *characteristic function* of the original game, and the resulting game in coalition form is called a game in *characteristic function form*. Examples of calculations of $v(S)$ were given in Example 4.30. Note that $v(\emptyset) = 0$. Also, $v(N)$ is the largest sum of the pay-offs to all the players, taken with respect to all possible choices of strategies. To find $v(N)$, just look at the original pay-off matrix, and locate the entry for which the sum of the coordinates is maximal. That sum is $v(N)$. For example, for the matrix game of Exercise 4.32(a), $v(N) = v(\{A,B,C\}) = 2$ and this is attained at (A_2, B_1, C_1). For the matrix game of Exercise 4.30(a), $v(N) = 7$, attained at (A_1, B_2, C_1) and at (A_2, B_1, C_2).

Exercise 4.33
Calculate the characteristic functions for the matrix games of Exercise 4.32 (a to c).

Note. The presentation of a game in characteristic function form sometimes obliterates important features of the game. This is illustrated by the following example given in McKinsey [1952], p. 351. Consider a two-person game in which player A has only one strategy and player B has two strategies, with the following game matrix:

$$
\begin{array}{ccc}
 & B_1 & B_2 \\
A_1 & (0, -1000) & (10, 0)
\end{array}
$$

Strategy B_2 dominates B_1 and, therefore, A will get a pay-off of 10 and B will receive 0. However, in characteristic function form, $v(A) = v(B) = 0$ and $v(\{A,B\}) = 10$, so that A and B seem to have the same strength.[20]

Theorem 4.2
A game in characteristic function form has the following *superadditivity property*:

$$v(S \cup T) \geq v(S) + v(T) \text{ for all coalitions } S \text{ and } T \text{ such that } S \cap T = \emptyset.$$

Proof. Let $\mathscr{S}_1$ be a choice of strategies for the players in S such that the sum of the pay-offs to the players in S when they play against any choice of strategies for players in $N - S$ is at least $v(S)$. Similarly, let $\mathscr{T}_1$ be a choice of strategies for the players in T such that the sum of the pay-offs to the players in T when they play against any choice of strategies for players in $N - T$ is at least $v(T)$. Then $\mathscr{S}_1 \cup \mathscr{T}_1$ is a choice of strategies for the players in $S \cup T$ such that the sum of the pay-offs for the players in $S \cup T$ is at least $v(S) + v(T)$ when they play against any choice of strategies for players in $N - (S \cup T)$. Hence, $v(S \cup T) \geq v(S) + v(T)$. ∎

---

[20]See the discussion in Luce-Raiffa [1957], Section 8.5.

Notation

For a game (N,v) in coalition form, where $N = \{a_1,\ldots, a_n\}$, let v_j stand for $v(\{a_j\})$ for $1 \leq j \leq n$.

Exercise 4.34

For a game (N,v) in coalition form, if the superadditivity property holds, show that $v(S) \geq \sum_{a_j \in S} v_j$.

Example 4.32

Let N be any nonempty finite set, and, for any subset S of N, let $v(S)$ be the number of members of S. Then (N,v) is a game in coalition form for which the superadditivity property holds. Note that the following property also holds:

(**Monotonicity**) If $S \subseteq T$, then $v(S) \leq v(T)$, for any coalitions S and T.

Definition

By a $\{0,1\}$ *coalition game*, we mean a game (N,v) in coalition form such that, for any coalition S, $v(S)$ is either 0 or 1. For such a game, a coalition S such that $v(S) = 1$ will be called a *winning coalition*.

Example 4.33

Let N be any nonempty set, let $v(S) = 1$ for any nonempty subset S of N, and let $v(\emptyset) = 0$. Then (N,v) is a $\{0,1\}$ game in coalition form. Assume now that N contains at least two distinct elements b and c. Let $S = \{b\}$ and $T = \{c\}$. Then $v(S \cup T) = 1$, $v(S) = 1$, and $v(T) = 1$. Hence, the superadditivity property does not hold in this game in coalition form. It is easy to see that the monotonicity property does hold.

Example 4.34

Let N be any nonempty finite set and let $v(S) = 1$ for any subset S that contains a majority (that is, more than half) of N, and $v(S) = 0$ for all other subsets S. It is easy to see that this $\{0,1\}$ coalition game satisfies both the superadditivity and the monotonicity property.

Example 4.35

By a *k-person weighted majority game with quota q* we mean a $\{0,1\}$ coalition game based on a finite nonempty set N and an assignment of nonnegative real numbers w_x for each x in N; a winning coalition S is any subset of N such that $\sum_{x \in S} w_x > q$.[21] The number w_x is called the *weight* of x.

Coalition games have a role to play in the analysis of the political power of individuals in a legislature, committee, or similar group. Let (N,v) be a

---

[21]In some examples, it may be more appropriate to use $\geq$ instead of $>$.

game in coalition form. For any b and c in N, we shall say that b and c are *mutually substitutable* with respect to (N,v) if, for all coalitions S such that b and c are not in S, $v(S \cup \{b\}) = v(S \cup \{c\})$. (In particular, $v(\{b\}) = v(\{c\})$.) A member d of N is called a *null-player* in (N,v) if, for all coalitions S, $v(S \cup \{d\}) = v(S)$. Hence, using the case where $S = \varnothing$, we infer that $v(\{d\}) = 0$ for any null-player d.

4.9   The Shapley value

Let $N = \{a_1,\ldots, a_n\}$, and let $G(N)$ be the set of all coalition games of the form (N,v). As usual, let R^n stand for n-dimensional Euclidean space, that is, the set of all n-tuples of real numbers.

Definition
By a Shapley value on N, we mean a function Φ from $G(N)$ into R^n that satisfies the following four conditions for any $g = (N,v)$ in $G(N)$:

1. (Nullity) If a_j is a null-player in g, then $(\Phi(g))_j = 0$.[22]
2. (Efficiency) $\sum_{j=1}^{n} (\Phi(g))_j = v(N)$
3. (Additivity) If $h = (N,v^*)$ is any game in $G(N)$, then $(\Phi(g + h))_j = (\Phi(g))_j + (\Phi(h))_j$. (Here, $g + h$ is the coalition game that assigns to any coalition S the value $v(S) + v^*(S)$.)
4. (Symmetry) If a_j and a_k are mutually substitutable with respect to g, then $(\Phi(g))_j = (\Phi(g))_k$.

An intended interpretation of a Shapley value is that it assigns to each player in a game (N,v) a measure of the "power" of that player in the game. Alternatively, the Shapley value may be considered as a reasonable assignment of pay-offs to the players. Some subsequent examples may make this interpretation clearer.

Theorem 4.3
(Shapley's Theorem) (Shapley [1953]) The following function Φ is the unique Shapley value on (N,v), where $N = \{a_1,\ldots,a_n\}$. For any game $g = (N,v)$ in $G(N)$

$$\textbf{(SF)} \qquad (\Phi(g))_j = \frac{1}{n!} \sum_r (v(S_j \cup \{a_j\}) - v(S_j))$$

where the summation is taken with respect to all $n!$ linear orderings r of N, and S_j denotes the set of players preceding a_j in the linear ordering r.

We shall postpone the proof of Shapley's Theorem (see Exercise 4.41) in order to present examples of the application of Shapley's formula (SF). One

---

[22]$(\Phi(g))_j$ denotes the jth coordinate of the n-tuple $\Phi(g)$.

way of looking at (SF) is to imagine successive coalitions being built up one element at a time in a random way, until we obtain all of N. We take the marginal contribution $v(S_j \cup \{a_j\}) - v(S_j)$ that a_j makes in this construction, and $(\Phi(g))_j$ is set equal to the average of all those contributions.

Example 4.35
(Majority vote) Let $N = \{a_1, a_2, a_3\}$ and let (N,v) be the $\{0,1\}$ coalition game in which the winning coalitions consist of a majority of the membership, that is, $\{a_1,a_2\}$, $\{a_1,a_3\}$, $\{a_2,a_3\}$, $\{a_1,a_2,a_3\}$. Let us find the Shapley value of a_3. The linear order $a_1 < a_2 < a_3$ yields a contribution to the sum of $v(\{a_1, a_2, a_3\}) - v(\{a_1,a_2\}) = 1 - 1 = 0$. The linear order $a_1 < a_3 < a_2$ produces $v(\{a_1,a_3\}) - v(\{a_1\}) = 1 - 0 = 1$. The linear order $a_2 < a_1 < a_3$ yields $v(\{a_1,a_2,a_3\}) - v(\{a_1,a_2\}) = 1 - 1 = 0$. The linear order $a_2 < a_3 < a_1$ yields $v(\{a_2,a_3\}) - v(\{a_2\}) = 1 - 0 = 1$. The linear orders $a_3 < a_1 < a_2$ and $a_3 < a_2 < a_1$ both yield 0. Hence, by (SF), $(\Phi(g))_3 = \frac{1}{3!}$ (2) = $^1/_3$. By symmetry, the computations of the Shapley values for a_1 and a_2 also yield $^1/_3$.

Exercise 4.35
1. Let $N = \{a_1, a_2, a_3, a_4\}$ and let (N,v) be the $\{0,1\}$ coalition game in which the winning coalitions consist of a majority of the membership, that is, $\{a_1,a_2,a_3\}$, $\{a_1,a_2,a_4\}$, $\{a_1,a_3,a_4\}$, $\{a_2,a_3,a_4\}$, $\{a_1,a_2,a_3,a_4\}$. Show that the Shapley value for each a_j in N is $^1/_4$.
2. Let $N = \{a_1, a_2, \dots, a_n\}$ and let (N,v) be the $\{0, 1\}$ coalition game in which the winning coalitions consist of a majority of the membership. Show that the Shapley value for each member a_j in N is $\frac{1}{n}$.

Example 4.36
(Market with one seller and two buyers) Let $N = \{a_1, a_2, a_3\}$ and let (N,v) be the $\{0,1\}$ coalition game in which the winning coalitions must contain the seller a_1 and at least one of the buyers a_2 and a_3. (Thus, "success" occurs when a sale can take place.) Let us use the Shapley formula (SF) to find the Shapley value of the seller a_1. The linear orders $a_2 < a_1 < a_3$, $a_3 < a_1 < a_2$, $a_2 < a_3 < a_1$, $a_3 < a_2 < a_1$ all contribute 1 to the sum, whereas the linear orders $a_1 < a_2 < a_3$ and $a_1 < a_3 < a_2$ contribute 0. Hence, the Shapley value of a_1 is $\frac{1}{3!}$ (4) = $^2/_3$. The Shapley value of a_2 is $\frac{1}{6}$, since the only nonzero contribution 1 to the sum comes from the linear order $a_1 < a_2 < a_3$. Likewise, the Shapley value of a_3 is $\frac{1}{6}$.

Exercise 4.36
1. (Market with one seller and three buyers) Let $N = \{a_1,a_2,a_3,a_4\}$ and let (N,v) be the $\{0,1\}$ coalition game with one seller a_1 and three buyers a_2, a_3, and a_4, that is, the winning coalitions must contain a_1 and at least one more member. Show that the Shapley value of a_1 is $^3/_4$ and the other Shapley values are $1/12$.
2. (Market with one seller and k buyers) Let $N = \{a_1, a_2, \dots, a_{k+1}\}$ and let (N,v) be the $\{0,1\}$ coalition game in which the winning coalitions must

contain a_1 and at least one more member. Find formulas for the Shapley value of a_1 and the Shapley value of the other players.

Example 4.37

(Political parties) Assume that there are five political parties, labeled p_1, p_2, p_3, p_4, p_5, and that one-third of the voting public supports party p_1, while the support of the rest is divided evenly among the other four parties. We can think of an election as a weighted majority game with quota $^1/_2$, where the weight w_1 assigned to p_1 is 1/3 and each of the other parties is assigned weight 1/6. The Shapley values assigned to the players in this game can be considered to be measures of the relative power of the parties. (In such political contexts, the Shapley value is also called the Shapley-Shubik power index. See Shapley-Shubik [1954].) Let us see what value is assigned to party p_1 by the Shapley formula (SF). For each of the 4! linear orderings $a < b < p_1 < c < d$ in which p_1 occurs third, there is a contribution of 1 to the sum. (Note that the sum of the weights of the parties preceding p_1 is 1/3 and, when the weight of p_1 is then added, the total becomes 2/3, which is greater than $^1/_2$.) A similar result holds for the linear orderings $a < b < c < p_1 < d$ in which p_1 occurs fourth. (The sum of the weights of the parties preceding p_1 is $^1/_2$, which is not greater than $^1/_2$, and then, when the weight of p_1 is added in, the total becomes 5/6, which is greater than $^1/_2$.) There is no contribution from any other linear orderings. (For linear orderings $a < p_1 < b < c < d$ in which p_1 occurs second, the sum of the weights of a and p_1 is $^1/_2$, which is not greater than $^1/_2$. For linear ordering $a < b < c < d < p_1$ in which p_1 occurs last, the sum of the weights of the four parties preceding p_1 is 2/3, which is greater than $^1/_2$.) So, the value of p_1 given by (SF) is $\frac{1}{5!}$ (4! + 4!) = 2/5. We leave it to the reader to show that each of the four other parties has the Shapley value 3/20. Notice that this is one-fourth of $(1 - (2/5))$.

Exercise 4.37

1. In a vote on a bill in the United States Senate, a winning coalition consists of a majority of the one hundred Senators or, in case of a tie vote, half of the Senate plus the Vice-President of the United States. What power is assigned by the Shapley value to the individual Senators and to the Vice-President?

2. Assume that a committee consists of $2k$ members, plus a non-voting chairperson. A motion passes either by winning a majority of the $2k$ members or, if there is a tie, by gaining the approval of the chairperson. What power does the Shapley value attribute to the ordinary members and to the chairperson?

Exercise 4.38

1. In a committee of six members, let a winning coalition on a vote consist either of a majority of the committee or, in case of a tie, the side that contains the chairperson of the committee (who is a member of the committee). Show that the Shapley value of the chairman is 1/3 and that the other members receive a value of 2/15.

2. Under the same voting rule as in Part (1), what are the Shapley values of the chaiperson and the other members when the total committee membership is $2k$?

Exercise 4.39

1. Consider the $\{0, 1\}$ coalition game on $N = \{a_1, a_2, a_3\}$ in which the only winning coalitions are $\{a_1, a_2, a_3\}$ and $\{a_1, a_2\}$. Find the Shapley values of a_1, a_2, and a_3.
2. Consider the coalition game (N,v) with $N = \{a_1, a_2, a_3\}$, $v(\{a_1, a_2, a_3\}) = v(\{a_1, a_2\}) = 5$, and $v(S) = 0$ for all other coalitions S. Find the Shapley values of a_1, a_2, and a_3.
3. For the following weighted majority games, find the Shapley value of the players. (Recall that a winning coalition is one for which the sum of the weights of its members is greater than the quota q.[23]) (i) Three players. $q = 2.5$, players A, B, C have weights 2, 1, 1. (ii) Three players. $q = 2.5$, players A, B, C have weights 2, 2, 1. (iii) Four players. $q = 3.5$, players A, B, C, D have weights 3, 2, 2, 1. (iv) Four players. $q = 6.5$, players A, B, C, D have weights 5, 3, 2, 1. (v) Three players. $q = .5$, players A, B, C have weights 15, 15, 1. (vi) Three players. $q = .5$, players A, B, C have weights 16, 15, 1 (vii) Twenty players. $q = .5$, $w_1 = 15$, all other w_j's are 1.

Exercise 4.40

(a) Fill in the details of the following proof that the function given by Shapley's formula (SF) satisfies the four conditions of the definition of a Shapley value.[24]

1. If a_j is a null-player, then each term $v(S_j \cup \{a_j\}) - v(S_j)$ in the formula (SF) is 0 and, therefore, $(\Phi(g))_j = 0$.
2. Consider a particular linear ordering r of N, say, $b_1 < b_2 < \cdots < b_n$. In $\sum_{j=1}^{n} (\Phi(g))_j$, consider the sum of the terms $v(S_j \cup \{a_j\}) - v(S_j)$ that come from r in each $(\Phi(g))_j$:

$$[v(\{b_1\}) - v(\varnothing)] + [v(\{b_1, b_2\}) - v(\{b_1\})] + [v(\{b_1, b_2, b_3\}) - v(\{b_1, b_2\})] + \cdots +$$

$$[v(\{b_1, ..., b_{n-1}\}) - v(\{b_1, ..., b_{n-2}\})] + [v(\{b_1, ..., b_n\}) - v(\{b_1, ..., b_{n-1}\})]$$

This sum reduces to $v(N)$, since $v(\varnothing) = 0$ and $\{b_1, ..., b_n\} = N$. Since there are $n!$ linear orderings of N, we get a total of $n!$ $v(N)$. But this is multiplied by $\frac{1}{n!}$ in (SF), and we wind up with $v(N)$.
3. The additivity property follows immediately from the definitions.
4. To show the symmetry property, assume a_j and a_k are mutually substitutable. Then it is easy to see that, for every term $v(S_j \cup \{a_j\}) - v(S_j)$, corresponding to a linear ordering r in the (SF) sum for $(\Phi(g))_j$, there is an equal term in the (SF) sum for $(\Phi(g))_k$, and vice versa. In fact,

---

[23]Note that the sum of the weights must be strictly greater than q. In some treatments, the sum of the weights is required only to be greater than or equal to q.
[24]The arguments in parts (a) and (b) of this exercise are based on Aumann [1989], pp. 30–32.

take the term in the (SF) sum for $(\Phi(g))_k$ that corresponds to the linear order obtained from r by switching a_j and a_k. It is easy to verify that the terms are equal, whether or not a_j precedes a_k in r.

(b) Fill in the details of the following sketch of a proof that there is a unique Shapley value on a nonempty finite set $N = \{a_1,\ldots, a_n\}$. Let f be a Shapley value on N. For every nonempty coalition T, let (N, v_T) be the $\{0,1\}$ coalition game (denoted simply by v_T) whose winning coalitions have T as a subset. For any real number c, the elements of $N - T$ are null-players in the game cv_T and any two elements of T are mutually substitutable in cv_T. By the nullity condition, $f(cv_T)_j = 0$ for j in $N - T$. By the symmetry condition, $f(cv_T)_j = f(cv_T)_k$ for j and k in T. Therefore, by the Efficiency condition, $\sum_{j=1}^{n} f(cv_T)_j = (cv_T)(N) = c(v_T(N)) = c$. Hence, $c = \sum_{j \in T} f(cv_T)_j = |T| f(cv_T)_j$ for every j in T, where $|T|$ denotes the number of elements in T. Thus, $f(cv_T)_j$ is equal to $c/|T|$ for j in T and is equal to 0 for j in $N - T$. But, $G(N)$ is the Euclidean space of dimension $m = 2^n - 1$ and there are m games v_T. We have determined $f(cv_T)$ for arbitrary c and T and, consequently, also $f(\sum_{i=1}^{m} c_i v_{T_i})$ for linear combinations $\sum_{i=1}^{m} c_i v_{T_i}$ of the v_T's. If we can show that the v_T's are linearly independent, then those linear combinations would make up all of $G(N)$ and the Shapley value would be uniquely determined. So, assume that the v_T's are linearly dependent. Then we could obtain an equation $v_T = \sum_{i=1}^{k} d_i v_{T_i}$ with $|T| \leq |T_i|$ for all i and all the T_i's different from each other and from T. Then $1 = v_T(T) = \sum_{i=1}^{k} d_i v_{T_i}(T) = \sum_{i=1}^{k} d_i(0) = 0$, which is impossible. Hence the v_T's are linearly independent.

The Shapley value is not the only imaginable way to estimate the power of individuals in legislatures, committees, or other political groups. The Banzhaf power index (Banzhaf [1965]) is just one of various alternative methods (cf. Lucas [1983], Straffin [1980, 1983]).

4.10 *Imputations*

In a coalition game (N,v), one of the most important aspects of a game, the distribution of pay-offs to the individual players, is not clearly determined. Let $N = \{a_1,\ldots, a_n\}$ be the set of players and let v_j stand for $v(\{a_j\})$, the value of the coalition consisting of player a_j alone. We shall be concerned with possible pay-off vectors $X = (x_1,\ldots,x_n)$, where x_j is a real number that is the pay-off to player a_j.

Definitions

X is said to be *individually rational* if $x_j \geq v_j$ for $1 \leq j \leq n$. X is said to be *group rational* if $\sum_{j=1}^{n} x_j = v(N)$. X is an *imputation* if X is both individually and group rational.[25] For X to be "optimal" for the players, it is reasonable to require individual rationality, so that each player would get at least as

---

[25]The concept of *imputation* was introduced as a central idea in von Neumann–Morgenstern [1944].

much as his value when he plays alone. As for group rationality, note that, if $\sum_{j=1}^{n} x_j$ were less than $v(N)$, then the players could do better by playing in the coalition N and dividing among themselves the surplus $v(N) - \sum_{j=1}^{n} x_j$.

Exercise 4.41
Show that examples of imputations are given by the Shapley values $\Phi(g)$ for any game $g = (N,v)$, where v is superadditive. (Hint: The superadditivity is needed to prove individual rationality, whereas group rationality is simply the Efficiency property.)

The following class of imputations that might serve as plausible equilibrium points has turned out to be of significance in economics.

Definition
By the core[26] of a coalition game (N,v) we mean the set of all imputations $X = (x_1,\ldots, x_n)$ such that, for all coalitions S, $v(S) \leq \sum_{a_j \in S} x_j$. Membership of an imputation X in the core indicates an "efficient" assignment of pay-offs in the sense that, for any set S of players, the total sum $\sum_{a_j \in S} x_j$ assigned to the players in S is not less than the amount $v(S)$ they can get from the game by joining together in a coalition.

Example 4.38
Consider the three-person majority game (N,v): $N = \{a_1,a_2,a_3\}$, $v(\{a_1,a_2,a_3\}) = v(\{a_1,a_2\}) = v(\{a_1,a_3\}) = v(\{a_2,a_3\}) = 1$ and $v(S) = 0$ otherwise. Note that $v(N) = 1$ and $v_1 = v_2 = v_3 = 0$. So the imputations are the triples $X = (x_1,x_2,x_3)$ such that $x_1 \geq 0$, $x_2 \geq 0$, $x_3 \geq 0$, and $x_1 + x_2 + x_3 = 1$. (They form a solid triangle in the first octant, connecting the vertices $(1,0,0)$, $(0,1,0)$, and $(0,0,1)$.) To be in the core, we must also have $x_1 + x_2 \geq 1$, $x_1 + x_3 \geq 1$, $x_2 + x_3 \geq 1$. But, since $x_1 + x_2 + x_3 = 1$ implies $x_1 + x_2 \leq 1$, $x_1 + x_3 \leq 1$, and $x_2 + x_3 \leq 1$, it follows that $x_1 + x_2 = 1$, $x_1 + x_3 = 1$, $x_2 + x_3 = 1$. These entail $x_1 = x_2 = x_3$, which, together with $x_1 + x_2 + x_3 = 1$, yields $x_1 = x_2 = x_3 = \frac{1}{3}$. Thus, $1 = x_1 + x_2 = \frac{2}{3}$, which is impossible. Hence the core is empty.

Example 4.39
Consider the game of a market with one seller a_1 and two buyers, a_2 and a_3. (See Example 4.36.) Then $v(N) = v(\{a_1,a_2\}) = v(\{a_1,a_3\}) = 1$ and $v(S) = 0$ otherwise. The imputations are the triples (x_1,x_2,x_3) such that $x_1 \geq 0$, $x_2 \geq 0$, $x_3 \geq 0$, and $x_1 + x_2 + x_3 = 1$. For the core we also must have $x_1 + x_2 \geq 1$ and $x_1 + x_3 \geq 1$. From $x_1 + x_2 + x_3 = 1$ we also know that $x_1 + x_2 \leq 1$ and x_1

---

[26]The notion of the core seems to have been first introduced and studied by Gillies and Shapley in 1953. (See Gillies [1959]) It appeared essentially in economics in Edgeworth [1881] and Böhm-Bawerk [1891].

$+ x_3 \leq 1$. Therefore, $x_1 + x_2 = 1$ and $x_1 + x_3 = 1$. From $x_1 + x_2 = 1$ and $x_1 + x_2 + x_3 = 1$ we get $x_3 = 0$, and from $x_1 + x_3 = 1$ and $x_1 + x_2 + x_3 = 1$ we get $x_2 = 0$. From $x_1 + x_2 + x_3 = 1$ and $x_2 = x_3 = 0$ we infer $x_1 = 1$. Hence, the core consists of the triple $(1,0,0)$, yielding a pay-off of 1 to the seller a_1 and 0 to the two buyers a_2 and a_3.

Exercise 4.42
Find the core of the following coalition games (N,v).

1. $N = \{a_1, a_2\}$, $v(N) = 1$, and $v_1 = v_2 = 0$.
2. $N = \{a_1, a_2, a_3\}$, $v(N) = 1$, and $v(S) = 0$ otherwise.
3. The game of Example 4.32.
4. The games of Exercise 4.39(1), (2).

A coalition game (N,v) is said to be *inessential* if it satisfies the following additivity property: $v(S \cup T) = v(S) + v(T)$ for all coalitions S and T such that $S \cap T = \varnothing$. Otherwise, the game is called *essential*.

Exercise 4.43
1. Show that, if (N,v) is inessential, then $v(S) = \sum_{a_j \in S} v_j$ for any coalition S and, in particular, $v(N) = \sum_{j=1}^{n} v_j$. (Recall that, if $N = \{a_1, \ldots, a_n\}$, then v_j stands for $v(\{a_j\})$.)
2. If $v(N) = \sum_{j=1}^{n} v_j$ and v is superadditive, show that (N,v) is inessential.
3. If (N,v) is essential and v is superadditive, prove that (N,v) has infinitely many imputations.

Clearly, Exercise 4.43(a) shows that, in an inessential game, there is no motive for players to form a coalition, since they can obtain the same total pay-off by playing as individuals.

Exercise 4.44
1. Find the imputations and the core of an inessential game.
2. Determine the Shapley value of an inessential game.

Exercise 4.45
A game in coalition form (N,v) is said to be a *constant-sum game* if $v(S) + v(N - S) = v(N)$ for all coalitions S. If, in addition, $v(N) = 0$, (N,v) is called a *zero-sum game*.

1. Show that an inessential game must be constant-sum.
2. Show that, if a constant-sum game (N,v) is essential and superadditive, then its core is empty. (Hint: The essentiality implies that $\sum_{j=1}^{n} v_j < v(N)$. Assume X is in the core. Then $\sum_{j=2}^{n} x_j \geq v(\{a_2, \ldots, a_n\}) = v(N) - v_1$. Since X is an imputation, $x_1 + \sum_{j=2}^{n} x_j = v(N)$ and, therefore, $x_1 \leq$

v_1. Similarly, $x_j \leq v_j$ for all j. Hence, $\sum_{j=1}^{n} x_j \leq \sum_{j=1}^{n} v_j < v(N)$, contradicting the fact that X is an imputation.)

Exercise 4.46

Find out whether the games in characteristic function form determined by the following matrix games are essential. Indicate also whether they are constant-sum or zero-sum.

$$(a) \begin{pmatrix} (2,3) & (-1,2) \\ (1,0) & (0,1) \end{pmatrix} \qquad (b) \begin{pmatrix} (2,-1) & (0,1) \\ (1,0) & (3,-2) \end{pmatrix}$$

(c)

A_1	C_1	C_2	A_2	C_1	C_2
B_1	$(-3,1,2)$	$(2,-1,0)$	B_1	$(0,1,1)$	$(1,-1,1)$
B_2	$(1,0,1)$	$(0,3,-1)$	B_2	$(2,0,0)$	$(1,1,1)$

(d)

A_1	C_1	C_2	A_2	C_1	C_2
B_1	$(1,-1,0)$	$(0,0,0)$	B_1	$(0,2,-2)$	$(1,-2,1)$
B_2	$(-1,2,1)$	$(3,-3,0)$	B_2	$(1,0,-1)$	$(1,2,-3)$

$$(e) \begin{pmatrix} (-2,4) & (1,1) \\ (0,2) & (-1,3) \end{pmatrix}$$

Exercise 4.47

Show that the game in characteristic function form determined by a two-person zero-sum matrix game is inessential. (But compare the three-person zero-sum game in Exercise 4.46(d).)

Exercise 4.48

1. Prove that, if a matrix game is a zero-sum game, then the corresponding coalition game in characteristic function form is also a zero-sum game.
2. Prove the same result as in part (1), with zero-sum replaced by constant-sum.
3. Give an example to show that the converse of part (1) is not always true.

4.11   Strategic equivalence

Let us try to classify all coalition games (N,v) with $N = \{a_1, \dots, a_n\}$. We shall say that (N,u) is *strategically equivalent* to (N,v) if there are real numbers $d > 0$ and $e_1, \dots, e_n$ such that, for any coalition S, $u(S) = d(v(S)) + \sum_{a_j \in S} e_j$. (The basic idea is that d represents a change in unit of measurement, and the e_j's may be thought of as fixed rewards or penalties for each player.) Let us abbreviate "strategically equivalent" by "streq."

Exercise 4.49
Prove that strategic equivalence is an equivalence relation, that is,

1. Every coalition game is streq to itself
2. If (N,u) is streq to (N,v), then (N,v) is streq to (N,u)
3. If (N,u) is streq to (N,v) and (N,v) is streq to (N,w), then (N,u) is streq to (N,w).

Exercise 4.50
Assume (N,u) is streq to (N,v), where $u(S) = d(v(S)) + \sum_{a_j \in S} e_j$ for all coalitions S. Prove:

1. (N,u) is inessential if and only if (N,v) is inessential.
2. If $(x_1,\ldots, x_n)$ is an imputation in (N,v), then $(dx_1,\ldots, dx_n) + (e_1,\ldots, e_n)$ is an imputation in (N,u), and conversely.
3. An imputation $(x_1,\ldots, x_n)$ in (N,v) is in the core of (N,v) if and only if $(dx_1,\ldots, dx_n) + (e_1,\ldots, e_n)$ is in the core of (N,u).
4. A coalition game (N,v) is said to be *{0,1}-normalized* if $v(N) = 1$ and $v_j = v(\{a_j\}) = 0$ for $1 \le j \le n$. Then: (i) Every {0,1}-normalized game is essential. (ii) Every essential game is strategically equivalent to a {0,1}-normalized game.
5. Find {0,1}-normalized games that are strategically equivalent to the coalition games of Examples 4.29 and 4.30.
6. If two {0,1}-normalized games are strategically equivalent, show that they are the same game.
7. Prove that every constant-sum game is strategically equivalent to a zero-sum game.

For imputations $X = (x_1,\ldots, x_n)$ and $Y = (y_1,\ldots, y_n)$ in a game (N,v), with $N = \{a_1,\ldots, a_n\}$, let us say that X *dominates* Y *over a coalition* S (written $X >_S Y$) if (i) $x_j > y_j$ for all a_j in S and (ii) $v(S) \ge \sum_{a_j \in S} x_j$. Moreover, we say that X *dominates* Y if $X >_S Y$ for some nonempty coalition S.

Theorem 4.4
(a) For any imputation Y, if Y is in the core, then Y is not dominated by any imputation. (b) The converse of (a) holds if the game is superadditive. (In particular, the converse holds for games in characteristic function form.)

Proof

(a) Assume Y is in the core and some imputation X dominates Y over a nonempty coalition S. Then $x_j > y_j$ for all a_j in S. Since Y is in the core,

$v(S) \le \sum_{a_j \in S} y_j$. Hence, $v(S) < \sum_{a_j \in S} x_j$, contradicting clause (ii) of the fact that $\wedge >_S Y$.

(b) Assume Y is not in the core, that is, $v(S) > \sum_{a_j \in S} y_j$ for some coalition S.

Clearly, S must be nonempty. Let $K = \dfrac{1}{|S|} (v(S) - \sum_{a_j \in S} y_j)$. Note that $K > 0$. Define $X = (x_1, \dots, x_n)$ by setting $x_j = y_j + K$ when $a_j \in S$, and $x_j = v_j + \dfrac{1}{n - |S|} (v(N) - v(S) - \sum_{a_j \notin S} v_j)$ when $a_j \notin S$. By superadditivity, $v(N) - v(S) - \sum_{a_j \in S} v_j \ge 0$. So, $x_j \ge v_j$ for all j and $\sum_{j=1}^{n} x_j = v(N)$, whence X is an imputation. Also, $x_j > y_j$ for a_j in S and $v(S) = \sum_{a_j \in S} x_j$. Hence, X dominates Y over S. Thus, Y is dominated by some imputation.

Exercise 4.51
Show that the set of imputations and the core are convex and closed.

Exercise 4.52
Prove

(a) If $X >_S Y$ and $Y >_S Z$, then $X >_S Z$. (b) If $X >_S Y$, then it is not the case that $Y >_S X$.

Exercise 4.53
   1. Find a coalition game in which there are imputations X and Y such that X dominates Y and Y dominates X.
   2. Find a coalition game in which there are imputations X, Y, and Z such that X dominates Y and Y dominates Z, but X does not dominate Z.

4.12 Stable sets

In their pioneering work, Von Neumann-Morgenstern [1944], Section 30, offered a notion that might provide the best available "solution" to a coalition game. By a *stable set* they meant a set Γ of imputations satisfying the following two conditions:

   1. (Internal stability) No imputation in Γ is dominated by any other imputation in Γ.
   2. (External stability) If an imputation is not in Γ, then it is dominated by some imputation in Γ.

The intention behind the concept of a stable set Γ is that the imputations in Γ form a family of reasonable pay-offs to the players. Internal stability means that no member of Γ offers better pay-offs than another member of Γ for all players in a significant coalition, while external stability means that any imputation which is not dominated by any member of Γ at all must itself belong to Γ.

Note that the core must be a subset of any stable set. For, any imputation outside a stable set Γ must be dominated by an imputation in Γ, but an imputation in the core is not dominated by any imputation.

Example 4.40

Consider the three-person majority coalition game (Example 4.35). Here, $v(N) = v(\{a_1,a_2\}) = v(\{a_1,a_3\}) = v(\{a_2,a_3\}) = 1$ and $v_1 = v_2 = v_3 = 0$. Then $Z_1 = (0,1/2,1/2)$, $Z_2 = (1/2,0,1/2)$, $Z_3 = (1/2,1/2,0)$ are imputations that form a stable set Γ. It is not hard to verify that no member of Γ dominates any other member of Γ. (See Exercise 4.54.) Let us show that any imputation $Y = (b_1,b_2,b_3)$ not in Γ is dominated by some member of Γ. We know that each $b_j \geq 0$ and $b_1 + b_2 + b_3 = v(N) = 1$. Next, note that at least two b_j's are less than $1/2$. (Otherwise, at least two of the b_j's are $\geq 1/2$, say, $b_1 \geq 1/2$ and $b_2 \geq 1/2$. But then $b_1 = 1/2$ and $b_2 = 1/2$ and, therefore, $b_3 = 0$. Hence, $Y = Z_3$, contradicting the fact that Y is not in Γ.) Let us look at the case where $b_1 < 1/2$ and $b_2 < 1/2$. The other cases are similar. Then Y is dominated by Z_3 over the coalition $S = \{a_1,a_2\}$.

Exercise 4.54

Verify the details of the following argument that no member of Γ in Example 4.40 is dominated by any other member of Γ. For the sake of contradiction, let us assume that Z_1 dominates Z_2 over a nonempty coalition S. (The other cases are similar.) Then $S = \{a_2\}$. So, $v(S) = 0$ and the second coordinate of Z_1 is $1/2$, contradicting the second clause of the dominance requirement.

Exercise 4.55

In the three-person majority coalition game of Example 4.40, show that, if $0 \leq r < 1/2$, then the set Ω of all imputations (r,y,z) is another stable set.

Exercise 4.56

Consider the four-player $\{0,1\}$ coalition game (N,v) in which $v(S) = 1$ for just those coalitions S containing at least two elements.

1. Describe the set of imputations.
2. Show that the set Δ of all imputations $X = (x_1,x_2,0,0)$ is a stable set.

Exercise 4.57

In a coalition game, let *UND* denote the set of all imputations that are dominated by no imputations at all. Prove: (a) The core is a subset of *UND* and *UND* is a subset of any stable set. (b) If the game is superadditive, then *UND* and the core are the same. (c) In the three-person majority game of Example 4.40, *UND* is the empty set.

Exercise 4.58

Find a coalition game in which the core is a stable set.

One drawback of stable sets is that most games have more than one stable set. (We already have seen this in the case of the three-person majority game of Example 4.40.) A more serious drawback is that, contrary to early hopes, there are games that have no stable set at all. The first such example was given by Lucas [1964].[27] An interesting alternative to the theory of stable sets, but with its own advantages and drawbacks, may be found in the theory of bargaining sets in Aumann-Maschler [1964]. Other approaches are discussed in Luce-Raiffa [1957], Sections 9 to 11.

---

[27]In the example $(N,v) = (\{1, 2, \ldots, 10\}, v)$, $v(S) = 0$ for all coalitions except the following:
$v(\{1,2\}) = v(\{3,4\}) = v(\{5,6\}) = v(\{7,8\}) = v(\{9,10\}) = 1$, and $v(\{3,5,7\}) = v(\{1,5,7\}) = v(\{1,3,7\}) = v(\{3,5,9\}) = v(\{1,3,9\}) = v(\{1,5,9\}) = v(\{1,4,7,9\}) = v(\{3,6,7,9\}) = v(\{2,5,7,9\}) = 2$, and $v(\{1,3,7,9\}) = v(\{1,5,7,9\}) = v(\{3,5,7,9\}) = 3$, and $v(\{1,3,5,7,9\}) = 4$, and $v(N) = 5$. This has a nonempty core, but there is an example in Lucas-Rabie [1980] with fourteen players and an empty core.

appendix one

Finite probability theory

Probability spaces

A *probability space* is determined by two things, its sample space and its probability function. By a *sample space* we shall mean a finite set $S = \{O_1, O_2, \ldots, O_n\}$. The elements O_j are called *outcomes*. Each outcome O_j is assigned a real number $P(O_j)$, called the *probability* of O_j. The following two assumptions are made about the probability function:

1. $0 \leq P(O_j) \leq 1$ for all O_j
2. $\sum_{j=1}^{n} P(O_j) = 1$, that is, $P(O_1) + \cdots + P(O_n) = 1$.

The outcomes are to be thought of intuitively as the possible results of an experiment or observation.

Examples

1. Throw a "fair" coin, that is, a coin that seems to be normal and balanced. The possible results are taken to be Head or Tail. These are the "outcomes" of the sample space $S = \{H,T\}$. The probability assignment defines $P(H) = \frac{1}{2}$ and $P(T) = \frac{1}{2}$.
2. Throw a "fair" die. The outcomes are the numbers that appear on the faces of the die. So, $S = \{1, 2, 3, 4, 5, 6\}$ and we let $P(j) = 1/6$ for $1 \leq j \leq 6$.
3. Consider the case of a "biased" coin; say, $S = \{H, T\}$ and $P(H) = .1$ and $P(T) = .9$.

In most scientific experiments or observations it is customary to assume the *frequency interpretation* of probability assertions. If O_j is one of the possible outcomes, then $P(O_j)$ is to be understood in the following way. Let m be a positive integer. Repeat the experiment (or observation) m times. Then the outcome O_j will occur a certain number of times, say k times. By the *relative frequency* of O_j in that sequence of experiments, we mean the ratio k/m. Note that $0 \leq k/m \leq 1$. If this relative frequency approaches a limit as m approaches infinity, that is, as m increases without bound, then that limit is taken to be the probability $P(O_j)$ of the outcome O_j. It can be seen easily that assumptions (1) and (2) about the probability assignments are valid. For (1), note that, since each relative frequency k/m lies between 0 and 1, the limit $P(O_j)$ must lie between 0 and 1. For (2), note that, if k_j is the number of occurrence of outcome O_j in a sequence of m experiments, then $k_1 + \cdots + k_n = m$ and, therefore, $(k_1/m) + \cdots + (k_n/m) = 1$. Therefore, (2) follows when we take the limit as m approaches infinity.

For example, to say that the probability of obtaining a Head when we toss a particular coin is $1/2$ would mean that, if we continue to toss that coin, then the relative frequency of Heads will eventually get as close as we want to $1/2$ if we toss the coin sufficiently many times. Note that this assertion that $P(H) = 1/2$ is a factual assertion about that particular coin. It has nothing to do with the fact that getting a Head is one of two possible outcomes. If the coin is weighted in a special way, $P(H)$ might be different from $1/2$.

Events. By an *event* in a probability space, we mean any subset of the sample space S.

Examples
1. Throw a die. The sample space $S = \{1, 2, 3, 4, 5, 6\}$. The event E of getting an even number consists of the subset $\{2, 4, 6\}$. The event F of getting a number greater than 4 consists of the subset $\{5,6\}$.
2. Throw a pair of dice. The outcomes in the sample space S consist of all pairs (i,j), where i and j are numbers from 1 to 6. (We can distinguish between the two dice and call one of them "the first die" and the other "the second die". Then i designates the number showing on the first die and j designates the number on the second die.) There are 36 such pairs (i,j). Let E_1 be the event in which the sum on the dice is 7. Thus, E_1 is the subset $\{(1,6), (6,1), (2,5), (5,2), (3,4), (4,3)\}$. Let E_2 be the event that the sum is 12, called "boxcars" in dice slang. Then E_2 is the one-element subset $\{(6,6)\}$.
3. Toss a coin three times. The outcomes consist of all sequences of three H's and T's. Since each toss has two results (H or T), then the total number of outcomes is $2 \times 2 \times 2 = 8$. In fact, $S = \{(H,H,H), (H,H,T), (H,T,H), (H,T,T), (T,H,H), (T,H,T), (T,T,H), (T,T,T)\}$. The event E_1 that there are more Heads than Tails is the subset $\{(H,H,H),(H,H,T), (H,T,H),(T,H,H)\}$. The event E_2 that there are the same number of Heads as Tails is the empty subset $\emptyset$. The event E_3 that the first toss is a Head is the subset $\{(H,H,H), (H,H,T), (H,T,H), (H,T,T)\}$.

Notice that an event can be specified by listing the outcomes that belong to the event or by a description of those outcomes in ordinary English. Given an event E of a probability space, we define the *probability* $P(E)$ as the sum of the probabilities of all the outcomes in the event E. For instance, in example

(1) above, where the event E is that an even number appears when a die is thrown, $P(E) = P(2) + P(4) + P(6)$. When F is the event that a number greater than 4 appears, $P(F) = P(5) + P(6)$.

Notice that no trouble arises because we are using the same letter "P" for the probability of an outcome and the probability of an event. In the very special case when an event E consists of a single outcome O_j, that is, when $E = \{O_j\}$, $P(E) = P(O_j)$ and there is no possibility of confusion.

The equiprobable case

It happens often that, in a particular probability space, all of the outcomes have the same probability. When this occurs, we say that we are dealing with the *equiprobable case*.

Definition
The equiprobable case occurs when $P(O_j) = P(O_k)$ for all outcomes O_j and O_k. There are certain phrases that indicate that the equiprobable case is intended. One of these is the word "fair". A fair die is one for which all six outcomes have the same probability 1/6. When a pair of dice is said to be fair, this means that all 36 outcomes have the same probability, 1/36. The phrase "at random" always signifies that the equiprobable case holds. If a card is picked at random from a deck, this means that all of the 52 possible outcomes have the same probability, 1/52.

Theorem A1.1
If the equiprobable case applies to a probability space having n elements in its sample space, then the probability of each outcome is $1/n$.

Proof. Let $S = \{O_1, \ldots, O_n\}$. Let c be the probability of each outcome. Since $P(O_1) + \cdots + P(O_n) = 1$, we have $c + \cdots + c = 1$, that is, $nc = 1$ and, therefore, $c = 1/n$. ∎

Theorem A1.2
Let E be an event in a probability space for which the equiprobable case holds. If there are n outcomes in the whole sample space and there are k outcomes in the event E, then $P(E) = k/n$. In such a situation, k is said to be the number of *favorable cases*. Thus,

$$P(E) = \frac{number\ of\ favorable\ outcomes}{total\ number\ of\ possible\ outcomes}$$

Proof. This follows immediately from Theorem A1.1.

Examples

1. Throw a pair of fair dice. Let E be the event that the sum is 7 and let F be the event that the sum is 11. By Theorem A1.2, $P(E) = 6/36 = 1/6$ and $P(F) = 2/36 = 1/18$. Also, $P(E \text{ or } F) = 8/36 = 2/9$.
2. Toss a fair coin five times. There are $2^5 = 32$ possible outcomes. If E is the event that there are more Heads than Tails, there are 16 favorable outcomes, since the number of outcomes with more Heads than Tails is equal to the number of outcomes with more Tails than Heads (and there are no outcomes with the same number of Heads and Tails). Hence, $P(E) = 16/32 = {}^1/_2$. Let F be the event that the first toss is a Head. There are 16 outcomes in F. So, $P(F) = 16/32 = {}^1/_2$.
3. Pick a card at random from a deck. Let E be the event that the card is a Heart. Then $P(E) = 13/52 = {}^1/_4$. Let F be the event that the card is an Ace. Then $P(F) = 4/52 = 1/13$.

Theorem A1.3
Probability Laws

1. If E and F are disjoint events, that is, they have no outcomes in common, then $P(E \text{ or } F) = P(E) + P(F)$.
2. $P(\text{not } E) = 1 - P(E)$ and $P(E) = 1 - P(\text{not } E)$.
3. $P(E \text{ or } F) = P(E) + P(F) - P(E \text{ \& } F)$.

Proof
1. Let j be the number of outcomes in E and let k be the number of outcomes in F. Since E and F are disjoint, the number of outcomes in the event $(E \text{ or } F)$ is $j + k$. Hence, $P(E \text{ or } F) = (j + k)/n$, $P(E) = j/n$, and $P(F) = k/n$. Thus, $P(E \text{ or } F) = P(E) + P(F)$.
2. The event $(E \text{ or } (\text{not } E))$ consists of all possible outcomes and, therefore, by Theorem A1.2, $P(E \text{ or } (\text{not } E)) = 1$. Since E and $(\text{not } E)$ are disjoint, part (1) yields $1 = P(E \text{ or } (\text{not } E)) = P(E) + P(\text{not } E)$, from which (2) follows.
3. The event $(E \text{ or } F) = (E \text{ or } (F - E))$, where $F - E$ is the event consisting of all outcomes in F but not in E. Hence, by part (1), $P(E \text{ or } F) = P(E) + P(F - E)$, since E and $(F - E)$ are disjoint. Now, $F = ((F - E) \text{ or } (E \text{ \& } F))$ and, since $(F - E)$ and $(E \text{ \& } F)$ are disjoint, part (1) yields $P(F) = P(F - E) + P(E \text{ \& } F)$. So, $P(F - E) = P(F) - P(E \text{ \& } F)$. Thus, $P(E \text{ or } F) = P(E) + P(F - E) = P(E) + P(F) - P(E \text{ \& } F)$. ∎

Examples

1. Let a card be chosen at random from a deck. Let C be the event that the card is a Club and let D be the event that the card is a Diamond. Since C and D are disjoint, $P(C \text{ or } D) = P(C) + P(D) = (13/52) + (13/52) = 26/52 = {}^1/_2$. Let A be the event that the card is an Ace. Then, by Theorem A1.3(1), $P(C \text{ or } A) = P(C) + P(A) - P(A \text{ \& } C) = (13/52) + (4/52) - (1/52) = 16/52 = 4/13$.

2. Let a fair coin be tossed twice. Let E be the event that at least one of
    the tosses is a Head. There are 4 possible outcomes. The event (not
    E) occurs when both tosses are Tails, and this can occur in only one
    way. Hence, $P(\text{not } E) = {}^{1}/_{4}$ and, therefore, by Theorem A1.3(1), $P(E) =$
    $1 - P(\text{not } E) = 1 - {}^{1}/_{4} = {}^{3}/_{4}.$

Conditional probability

We are often interested in knowing the probability of a certain event E, given
that some other event F occurs. For example, if a fair coin is tossed five times,
what is the probability that more Heads than Tails appear, given that there
is at least one Tail ? The probability of E, given F, will be denoted $P(E/F)$. We
shall first define $P(E/F)$ and then motivate that definition.

Definition

$$P(E/F) = \frac{P(E \& F)}{P(F)}$$

To see why this definition is reasonable, assume that an experiment is
repeated m times and let k be the number occurrences of the event F.
Among these k occurrences of F, assume that j are also occurrences of
event E. Intuitively, j/k is the relative frequency of E with respect to F
and, therefore, $P(E/F)$ should be the limit of j/k as m approaches infinity.
But, $j/k = (j/m)/(k/m)$. Since k/m is the relative frequency of F and j/m is the
relative frequency of the event $E \& F$, $P(F)$ is the limit of k/m and $P(E \& F)$
is the limit of j/m as m approaches infinity. Thus, the limit of j/k is $P(E \& F)/$
$P(F)$, which we take to be the definition of $P(E/F)$. Note that $P(E/F)$ is not
defined when $P(F) = 0$.

Example

Toss a fair coin five times. Let E be the event that more Heads than Tails
appear and let F be the event that at least one Tail occurs. There are 32
possible outcomes. E occurs 16 times (half of the total number of occurrences,
since "more Tails than Heads" occurs as often as "more Heads than Tails").
Among those 16 occurrences of E, the event that there is no Tail occurs once;
hence, $E \& F$ occurs 15 times. So, $P(E \& F) = 15/32$. Note also that $P(F) = 1 -$
$P(\text{not } F) = 1 - (1/32) = 31/32$, since F occurs 31 times. Thus, $P(E/F) = P(E \& F)/$
$P(F) = (15/32)/(31/32) = 15/31.$

Theorem A1.4

$$P(E \ \& \ F) = P(F) \cdot P(E/F)$$

This follows from the definition of $P(E/F)$.

Example

Let two cards be chosen at random, one after the other, from an ordinary deck. The first card is not replaced before the second card is chosen. Let us find the probability that both cards are Kings. Let F be the event that the first card is a King and let E be the event that the second card is a King. Then E & F is the event that both cards are Kings. Now $P(E/F) = 3/51 = 1/17$, since, if the first card is a King, then there are 51 possible outcomes for the second card and 3 of them are Kings. By Theorem A1.4:

$$P(E \ \& \ F) = P(F) \cdot P(E/F) = (4/52) \cdot (1/17) = (1/13) \cdot (1/17) = 1/221.$$

(This also can be solved without using conditional probability. There are 52 × 51 outcomes. There are 4 × 3 ways of getting two Kings. Hence, the probability is $(4 \times 3)/(52 \times 51) = 1/221$.)

Exercise A1.1

1. If two fair dice are thrown, show that the probability that at least one die shows a 5, given that the dice show different numbers, is 1/3.
2. Assume that the probability that any child is a boy is $1/2$. In a family with two children, what is the probability that both children are boys, given that at least one is a boy? In a family with two children, what is the probability that both children are boys, given that the older child is a boy? In a family with three children, what is the probability that all three children are boys, given that at least one is a boy?

Exercise A1.2

(Bayes' Formula) Assume that there is a test for a certain disease that occurs in .1 percent of the population (that is, the probability that a person chosen at random has the disease is .001). Assume that the probability that a person who has the disease yields a positive result on the test is .98 and the probability that a person who does not have the disease yields a positive result on the test is .001. Find the probability that a person who gets a positive result on the test actually has the disease. (The experiment consists of testing an arbitrarily chosen person for the disease. Let D be the event that the person has the disease and let T be the event that the person gets a positive test result. Note that $P(D) = .001$, $P(T/D) = .98$, and $P(T/\text{not } D) = .001$. Observe also that $T = (T \ \& \ D)$ or $(T \ \& \ (\text{not } D))$. First find $P(T)$ and then $P(D/T)$.)

Independence

To say that two events E and F are independent means intuitively that knowing that one of them occurs gives us no new information about the probability of the other event. This amounts to saying that $P(E/F) = P(E)$ and $P(F/E) = P(F)$. This is equivalent to $P(E \& F)/P(F) = P(E)$ and $P(F \& E)/P(E) = P(F)$ and, therefore, simply to $P(E \& F) = P(E) \, P(F)$ (since $E \& F$ is the same event as $F \& E$). This inspires the following definition:

Definition
Events E and F are *independent* if and only if $P(E \& F) = P(E) \cdot P(F)$

Examples
Let two fair dice be thrown. Let E be the event that the sum of the dice is 8 and let F be the event that the first die is a 5. Are E and F independent?

$$P(F) = 1/6, \; P(E) = 5/36, \text{ and } P(E \& F) = 1/36$$

So,

$$P(E) \cdot P(F) = 5/216 \neq P(E \& F)$$

So, E and F are not independent. Now let G be the event that the sum of the dice is 7. Are G and F independent?

$$P(G) = 6/36 = 1/6$$

and

$$P(G \& F) = 1/36$$

Then

$$P(G) \cdot P(F) = 1/36 = P(G \& F)$$

So, G and F are independent.

Exercise A1.3
Assume that events E and F are independent. Show that E and (not F) also are independent. Show also that (not E) and (not F) are independent.

Exercise A1.4
Assume that the probability that a child is a boy is $1/2$. In a family, let E be the event that the family has children of both sexes, and let F be the event that the family has at most one girl. In a family with three children, are E and F independent? In a family with four children, are E and F independent?

Exercise A1.5
If a pair of fair dice is thrown and E is the event that the sum is even and F is the event that at least one die is a 3, are E and F independent?

Random variables

In a probability space, by a *random variable* we mean a function X that assigns to each outcome a real number.

Examples

1. Toss a fair coin five times. Let X be the number of Heads. Let Y be the number of Heads that are preceded by no Tails. X and Y are random variables.
2. Throw a pair of fair dice. Let X be the sum of the dice and let Y be the maximum of the numbers on the dice. X and Y are random variables.

By the *expected value* $E(X)$ of a random variable X, we have in mind the long-run average value of X as the underlying experiment is repeated. To be more specific, assume that the experiment occurs n times and that the values obtained for X are $u_1, \ldots, u_n$. Then the average value of X over that sequence of n experiments is $(u_1 + \cdots + u_n)/n$. If that average value approaches a limit as n approaches infinity, then that limit is denoted $E[X]$ and is called the *expected value* of X.

Let us try to use the frequency interpretation to suggest a formula for $E[X]$. Let $x_1, \ldots, x_k$ be the possible values of X. After n experiments, let m_i be the number of experiments in which X takes the value x_i. Then, as n increases, m_i/n approaches $P(X = x_i)$, and m_i is approximately $n\, P(X = x_i)$. So, when we compute the average value of X, the x_i terms in the numerator add up to approximately $n\, P(X = x_i)\, x_i$. Hence, the numerator is close to

$$\sum_{i=1}^{n} n\, P(X = x_i)\, x_i$$

So, dividing by n, we get an approximation to the average value of

$$\sum_{i=1}^{n} P(X = x_i)\, x_i$$

As n approaches infinity, the approximation gets better and better. Therefore, we take as our definition:

$$E[X] = \sum_{i=1}^{n} P(X = x_i)\, x_i$$

So, to find $E[X]$, we multiply each possible value of X by the probability that X has that value and then add up the results.

Example

Toss a fair coin 4 times. There are 16 outcomes. Let X be the number of Heads. The possible values of X are 0, 1, 2, 3, 4. Now, $P(X = 0) = 1/16$, $P(X = 1) = 4/16$, $P(X = 2) = 6/16$, $P(X = 3) = 4/16$, $P(X = 4) = 1/16$. Hence

$$E[X] = (0 \times (1/16)) + (1 \times (4/16)) + (2 \times (6/16)) + (3 \times (4/16)) + (4 \times (1/16))$$

$$= 32/16 = 2$$

Theorem A1.5

For any two random variables X and Y on the same probability space

$$E[X + Y] = E[X] + E[Y]$$

Proof. Let $x_1, \ldots, x_k$ and $y_1, \ldots, y_r$ be the possible values of X and Y, respectively.

$$\sum_{i,j} (x_i + y_j)\, P(X = x_i \ \& \ Y = y_j)$$

$$= \sum_{i,j} x_i\, P(X = x_i \ \& \ Y = y_j) + \sum_{i,j} y_j\, P(X = x_i \ \& \ Y = y_j)$$

$$= \sum_i \sum_j x_i\, P(X = x_i)\, P(Y = y_j / X = x_i) + \sum_j \sum_i y_j\, P(Y = y_j)P(X = x_i / Y = y_j)$$

$$= \sum_i x_i\, P(X = x_i) \sum_j P(Y = y_j / X = x_i) + \sum_j y_j\, P(Y = y_j) \sum_i P(X = x_i / Y = y_j)$$

$$= E[X] + E[Y].$$

At the last step, we have used the following computation:

$$\sum_j P(Y = y_j / X = x_i) = \sum_j P(Y = y_j \ \& \ X = x_i)/P(X = x_i)$$

$$= \left(\sum_j P(Y = y_j \ \& \ X = x_i) \right) / P(X = x_i)$$

$$= P((Y = y_1 \ \& \ X = x_i) \text{ or} \dots \text{or } (Y = y_r \ \& \ X = x_i))/P(X = x_i)$$

$$= P((Y = y_1 \text{ or} \dots \text{or } Y = y_r) \ \& \ X = x_i)/P(X = x_i) = P(X = x_i)/P(X = x_i) = 1.$$

A similar computation shows that

$$\sum_i P(X = x_i/Y = y_j) = 1. \qquad \blacksquare$$

Theorem A1.6

$$E[X_1 + \dots + X_n] = E[X_1] + \dots + E[X_n].$$

This follows from Theorem A1.5 by mathematical induction. When $n = 3$, the idea is made clear as follows:

$$E[X_1 + X_2 + X_3] = E[(X_1 + X_2) + X_3] = E[X_1 + X_2] + E[X_3]$$

$$= E[X_1] + E[X_2] + E[X_3]$$

Repeated Bernoulli trials

By a *Bernoulli trial* we mean an experiment with two possible outcomes, which we shall call "Success" and "Failure", and we denote the probability of Success by p and, the probability of Failure by $q = 1 - p$. Let us suppose that the experiment is repeated n times.

Let X_i be the number of successes on the ith trial. Thus, X_i has possible values 0 and 1, and $P(X_i = 1) = p$ and $P(X_i = 0) = 1 - p = q$. Therefore

$$E[X_i] = 0 \times P(X_i = 0) + 1 \times P(X_i = 1) = p$$

Now, if X is the total number of successes among the n trials, $X = X_1 + \dots + X_n$, and, by Theorem A1.6,

$$E[X] = E[X_1] + \dots + E[X_n] = p + \dots + p = np$$

So, we have the following result.

Theorem A1.7
The expected number of successes in n Bernoulli trials is np, where p is the probability of success.

Examples
   1. Toss a fair coin n times. By Theorem A1.7, the expected value of the number of Heads is $n/2$. Here, Success = Head, and $p = \frac{1}{2}$ because the coin is fair.

2. Throw a pair of fair dice 10 times. Then the expected number of times that the sum of the dice will be 7 is 5/3, since $n = 10$ and the probability p that the sum is 7 is $6/36 = 1/6$. By Theorem A1.7, the expected value is $np = 10(1/6) = 5/3$. It should not be surprising that the expected value is not an integer, since the expected value is the long-run average.

Exercise A1.6

1. If a random variable X has n values $x_1, \ldots, x_n$ and the probabilities $P(X = x_i)$ all have the same value (which must be $1/n$), show that $E[X]$ is the average $(x_1 + \cdots + x_n)/n$ of the possible values.
2. If X is the number shown when a fair die is thrown, show that $E[X] = 7/2$.
3. Show that the sum when a pair of fair dice is thrown has expected value 7.

Exercise A1.7

Random variables X and Y with possible values $x_1, \ldots, x_n$ and $y_1, \ldots, y_r$ respectively, are said to be *independent* if the events $X = x_i$ and $Y = y_j$ are independent for all x_i and y_j.

1. Show that, if X and Y are independent random variables, then $E[XY] = E[X] E[Y]$.
2. Give an example of two random variables X and Y for which $E[XY] \neq E[X] E[Y]$.

appendix two

Utility theory[1]

The numbers that are pay-offs of outcomes in a game often are given in terms of monetary units or other measures that may not adequately reflect the preferences of the players. Attempts to find a uniform measure for the various players may be difficult or impossible. But, for a single person, it is not unreasonable to expect that a numerical gauge of that person's values can be constructed.

Let us assume that there are finitely many outcomes that may occur, say, $A_1, \ldots, A_m$. These outcomes will be called *alternatives*. For any alternatives A_i and A_j, we denote by $A_i \gtrsim A_j$ that the player either prefers A_i to A_j or gives equal preference to A_i and A_j. We shall make certain assumptions, Axioms 1 to 6, about this relation.

Axiom 1
1. Connectedness. For any alternatives A_i and A_j, $A_i \gtrsim A_j$ or $A_j \gtrsim A_i$.
2. Transitivity. For any alternatives A_i, A_j, A_k, if $A_i \gtrsim A_j$ and $A_j \gtrsim A_k$, then $A_i \gtrsim A_k$.

As a consequence of Axiom 1 and by changing subscripts, we can assume the ordering

$$A_1 \gtrsim A_2 \gtrsim \ldots \gtrsim A_m$$

We shall assume that $A_m \not\gtrsim A_1$. Otherwise, all the alternatives would be preferred equally, which is possible, but certainly not interesting.[2] This additional hypothesis should be considered part of Axiom 1.

By a *lottery* for prizes $A_1, \ldots, A_m$, we mean an assignment of probabilities $p_1, \ldots, p_m$ to $A_1, \ldots, A_m$, respectively, such that $p_1 + \cdots + p_m = 1$. Such a lottery will be denoted $(p_1 A_1, \ldots, p_m A_m)$. We shall also deal with compound lotteries

---

[1]Our treatment is based upon the pioneering work in Von Neumann—Morgenstern [1944], Chapter 1, Section 3, and the Appendix, but the exposition follows more closely that in Luce-Raiffa [1957], Chapter 2.
[2]We say that A_i and A_j have *equal preference* and write $A_i \sim A_j$ if $A_i \gtrsim A_j$ and $A_j \gtrsim A_i$.

among lotteries. If $L_1, \ldots, L_n$ are lotteries, then, by the lottery $(q_1L_1, \ldots, q_nL_n)$ we mean the assignment of probabilities $q_1, \ldots, q_n$ to $L_1, \ldots, L_n$, respectively, such that $q_1 + \cdots q_n = 1$.

Axiom 2
The compound lottery $(q_1L_1, \ldots, q_nL_n)$ is preferred equally with the corresponding simple lottery for $A_1, \ldots, A_m$ that is defined in the expected way. In other words, if $L_i = (p_{i1}A_1, p_{i2}A_2, \ldots, p_{im}A_m)$ for $1 \le i \le n$, then the resulting lottery $(p_1A_1, \ldots, p_mA_m)$ is determined by the equations $p_j = q_1p_{1j} + q_2p_{2j} + \cdots + q_np_{nj}$ for $1 \le j \le m$.

Example
Assume that the alternatives are A_1, A_2, A_3, and that we are given two lotteries $L_1 = (\frac{1}{4}A_1, \frac{1}{2}A_2, \frac{1}{4}A_3)$ and $L_2 = (\frac{1}{8}A_1, \frac{3}{8}A_2, \frac{1}{2}A_3)$. Then the compound lottery $(\frac{1}{3}L_1, \frac{2}{3}L_2)$ is preferred equally with the lottery (p_1A_1, p_2A_2, p_3A_3), where

$$p_1 = \frac{1}{3}\frac{1}{4} + \frac{2}{3}\frac{1}{8} = 1/6$$

$$p_2 = \frac{1}{3}\frac{1}{2} + \frac{2}{3}\frac{3}{8} = 5/12$$

and

$$p_3 = \frac{1}{3}\frac{1}{4} + \frac{2}{3}\frac{1}{2} = 5/12.$$

In Axiom 2, it is assumed that the experiment yielding the compound lottery is independent of the experiments determining the lotteries $L_1, \ldots, L_n$. Note that, in Axiom 2, we have extended the notion of preference from alternatives to lotteries and compound lotteries.

Axiom 3   (Continuity)
Each alternative A_i is preferred equally with a suitable lottery involving only A_1 and A_m. (We shall use the notation: $A_i \sim (x_i A_1, (1-x_i) A_m)$. The latter lottery will be denoted by A_i^*. Note that $x_1 = 1$ and $x_m = 0$.)

The general validity of Axiom 3 has been questioned. For example, if $A_1 = \$10$, $A_2 = \$1$, and $A_3 = $ Death, is any lottery involving A_1 and A_3 of equal preference with A_2?

Axiom 4   (Substitutability)
In any lottery L, A_i is replaceable by A_i^*, that is, the new lottery has equal preference with the original one.

Axiom 5   (Transitivity)
Preference and equal preference are transitive.
We already know from Axiom 1 that this holds for alternatives.

Theorem 1
Any lottery is equally preferred with one involving only A_1 and A_m.

Proof. Given lottery $L = (p_1A_1, p_2A_2, \ldots, p_mA_m)$, replace each A_i by A_i^*. Recall that A_i^* is the lottery $(x_i A_1, (1 - x_i)A_m)$. By Substitutibility, $L \sim (p_1A_1^*, p_2A_2^*, \ldots, p_mA_m^*)$. By Axiom 2, $L \sim (p A_1, (1 - p) A_m)$, where $p = p_1x_1 + p_2x_2 + \cdots + p_mx_m$. ∎

Axiom 6   (Monotonicity)
The lottery $(p A_1, (1 - p) A_m) \succsim (q A_1, (1 - q) A_m)$ if and only if $p \geq q$.

Theorem 2
The order of preference between two lotteries $L_1 = (p_1A_1, p_2A_2, \ldots, p_mA_m)$ and $L_2 = (q_1A_1, q_2A_2, \ldots, q_mA_m)$ corresponds to the relation between the numbers $p_1x_1 + p_2x_2 + \cdots + p_mx_m$ and $q_1x_1 + q_2x_2 + \cdots + q_mx_m$

Proof
Just use the procedure in the proof of Theorem 1 and apply Axioms 5 and 6. ∎

By a *utility function* we mean a function u on the set $\mathcal{L}$ of lotteries determined by a set $A_1, \ldots, A_m$ of alternatives such that $u(L_1) \geq u(L_2)$ if and only if $L_1 \succsim L_2$. On the basis of Axioms 1 to 6, a utility function u can be defined such that:

$$u(A_1) = 1 = x_1$$

$$u(A_i) = x_i \quad \text{for } 1 < i < m$$

$$u(A_m) = 0 = x_m$$

$$u(p_1A_1, p_2A_2, \ldots, p_mA_m) = p_1x_1 + p_2x_2 + \cdots + p_mx_m.$$

A utility function u is said to be *linear* if $u(p L_1, (1 - p)L_2) = p\, u(L_1) + (1 - p)u(L_2)$ for any lotteries L_1 and L_2 and any number p such that $0 \leq p \leq 1$. The utility function just mentioned above can be shown to be linear. It is easy to see that, if u is a linear utility function and a and b are any numbers with $a > 0$, then the function $u^{\#}(L) = a\, u(L) + b$ is also a linear utility function. Conversely, if u^* is a linear utility function corresponding to the same preference relation as a linear utility function u, then it can be shown that there are real numbers a^* and b^*, with $a^* > 0$, such that $u^*(L) = a^*u(L) + b^*$ for all lotteries L. It follows that we always can replace a linear utility function by an equivalent one with 0 as least utility and 1 as greatest utility.

appendix three

Nash's theorem

Assume that a game has k players, each with a finite number of strategies. Nash's Theorem (see Nash [1950]) asserts that there is an equilibrium point, consisting of pure or mixed strategies. We shall sketch Nash's argument.

For $1 \leq i \leq k$, let $\boldsymbol{S}_i$ be the set of mixed strategies for the ith player. We shall be dealing with k-tuples $(S_1, \ldots, S_k)$, where each S_i is in $\boldsymbol{S}_i$.

Definition

Given two k-tuples $(S_1, \ldots, S_k)$ and $(T_1, \ldots, T_k)$ in the product space $\boldsymbol{S}_1 \times \cdots \times \boldsymbol{S}_k$, we say that $(S_1, \ldots, S_k)$ *counters* $(T_1, \ldots, T_k)$ if, for each S_j

$$P_j(T_1, T_2, \ldots, T_{j-1}, S_j, T_{j+1}, \ldots, T_k) \geq P_j(T_1, T_2, \ldots, T_{j-1}, S_j^*, T_{j+1}, \ldots, T_k)$$

for all S_j^* in $\boldsymbol{S}_j$. (Here, P_j is the pay-off function for the jth player.)

Theorem

$(S_1, S_2, \ldots, S_k)$ is an equilibrium point if and only if $(S_1, S_2, \ldots, S_k)$ counters itself.

This follows easily from the definition of "equilibrium point".

Define the following function G on $\boldsymbol{S}_1 \times \cdots \times \boldsymbol{S}_k$.
For any $(T_1, \ldots, T_k)$, let $G(T_1, \ldots, T_k)$ be the set of k-tuples that counter $(T_1, \ldots T_k)$. Note that, by the Theorem, $(S_1, \ldots, S_k)$ is an equilibrium point if and only if $(S_1, \ldots, S_k)$ belongs to $G(S_1, \ldots, S_k)$. It is not hard to show that the graph of G is closed (that is, if $p_1, p_2, \ldots$ and $q_1, q_2, \ldots$ are sequences of points in the product space such that p_i counters q_i for each i, and the sequences $\{p_i\}$ and $\{q_i\}$ converge to u and v, respectively, then u counters v). It is also fairly easy to show that each $G(T_1, \ldots, T_k)$ is convex. Then, by Kakutani's Fixed Point Theorem (in Kakutani [1941]), the function G has a fixed point (that is, a point p such that p belongs to $G(p)$). But a fixed point of G is an equilibrium point. This proves Nash's Theorem. ∎

A second proof, using Brouwer's Fixed Point Theorem

We shall carry out the proof for two-person games in order to make the notation less unwieldy. The proof for arbitrary k-person games is just an obvious generalization.[1]

Let $A_1, \ldots, A_m$ and $B_1, \ldots, B_n$ be the strategies for players A and B, respectively. Let $X = (x_1, \ldots, x_m)$ and $Y = (y_1, \ldots, y_n)$ be arbitrary mixed strategies for A and B, respectively. Define

$$r_i = \text{maximum } (P_A(A_i, Y) - P_A(X, Y), 0)$$

$$s_j = \text{maximum } (P_B(X, B_j) - P_B(X, Y), 0)$$

where P_A and P_B are the pay-off functions for A and B, respectively. Further, let

$$x_i^* = (x_i + r_i) / \left(1 + \sum_{i=1}^{m} r_i \right)$$

and

$$y_j^* = (y_j + s_j) / \left(1 + \sum_{j=1}^{n} s_j \right)$$

Let $X^* = (x_1^*, \ldots, x_m^*)$ and $Y^* = (y_1^*, \ldots, y_n^*)$. The function $F(X, Y) = (X^*, Y^*)$ is easily shown to be continuous.

1. X^* and Y^* are mixed strategies. (This is not difficult to show.)
2. For any X and Y, (X, Y) is an equilibrium pair if and only if $F(X, Y) = (X, Y)$.

Proof of (2). $\Rightarrow$ Assume (X, Y) is an equilibrium pair. Then $P_A(A_i, Y) \leq P_A(X, Y)$ for all i. Hence, $r_i = 0$ for all i and, therefore, $x_i^* = x_i$ for all i. Thus, $X^* = X$. Similarly, $Y^* = Y$.

$\Leftarrow$ Assume (X, Y) is not an equilibrium pair. Then either (*) there exists X' such that $P_A(X', Y) > P_A(X, Y)$ or (**) there exists Y' such that $P_B(X, Y') > P_B(X, Y)$. Say (*) holds. (A similar argument applies when (**) holds.) Since $P_A(X', Y) = \sum_{i=1}^{m} x_i' P_A(A_i, Y)$, there exists i such that $P_A(A_i, Y) > P_A(X, Y)$. (Otherwise, $P_A(X', Y) = \sum_{i=1}^{m} x_i' P_A(A_i, Y) \leq \sum_{i=1}^{m} x_i' P_A(X, Y) = P_A(X, Y) \sum_{i=1}^{m} x_i' = P_A(X, Y)$,

---

[1] Our proof is based on the one in Nash [1951].

contradicting (*).) Hence, $r_i > 0$ and, therefore, $\sum_{i=1}^{m} r_i > 0$. On the other hand, $P_A(X,Y) = \sum_{i=1}^{m} x_i \, P_A(A_i, Y)$. Therefore, $P_A(A_i, Y) \leq P_A(X,Y)$ for some i such that $x_i > 0$. Then, for such i, $r_i = 0$. So, $x_i^* = x_i/(1 + \sum_{i=1}^{m} r_i) < x_i$ and, therefore, $X^* \neq X$.

This completes the proof of (2). The set of all (X,Y) is a closed, bounded, convex subset of R^{m+n}. Since F is continuous, Brouwer's Fixed Point Theorem[2] implies that F must have a fixed point (X,Y). By (2), such an (X,Y) is an equilibrium pair. ∎

---

[2] The Brouwer Fixed Point Theorem says that, for any continuous function F from a closed, bounded, convex subset C of n-dimensional Euclidean space R^n into C, there exists a fixed point of F in C, namely, a point x in C such that $F(x) = x$.

Answers to selected exercises

Exercise 0.1
Games (1)–(5) all are finite, deterministic, zero-sum, with perfect information. Game (6) has all these properties except perfect information, and game (7) has all the properties except being deterministic. Game (8) is everything except deterministic and with perfect information. Game (9) has all the properties except being deterministic and finite. (A Head may never turn up, but the probability of this happening is 0.)

Exercise 0.2
3. A has a winning strategy. A should remove one stick. If B then removes one, A removes two and wins. If B removes two, A removes the last stick and wins.
4. B has a winning strategy. If A names an even integer, B should name an odd integer. If A names an odd integer, B should name an even integer.
5. B has a winning strategy. Whenever A names an integer x, B should name $10 - x$. After two moves, the sum is 10; after four moves, the sum is 20, and so on. After twenty moves, the sum is 100 and B wins.

Exercise 0.3
Both A and B have four strategies.

Exercise 1.1
All of them.

Exercise 1.2
1. B has a winning strategy: name the same integer that A named.
2. A has a winning strategy. A should name the integer 2.
3. B has a winning strategy. If A first removes one stick, B should remove two. If A first removes two sticks, B should remove one.

4. A has a winning strategy. A should first remove one stick. If B then removes two, then A should remove one. If B removes one stick, then A should remove two.
5. Consider the case when $n = 3k + 1$. B has a winning strategy. Whenever A removes x sticks (where x is 1 or 2), B should remove $3 - x$ sticks. Thus, three sticks are removed after every two moves. Eventually B leaves one stick, which A has to remove and lose. In the case when $n = 3k + 2$, A has a winning strategy. A should first remove one stick, leaving $3k + 1$ sticks and then A has the winning strategy just described. In the case when $n = 3k$, A again has a winning strategy. A should remove two sticks, leaving $3k - 2 = 3(k-1) + 1$ sticks and A then can play the winning strategy described above.

Exercise 1.3
A has a winning strategy when and only when $n = 3k + 1$ or $n = 3k + 2$. A should always leave a multiple of three sticks. If $n = 3k$, B can always leave a multiple of three sticks and, therefore, has a winning strategy.

Exercise 1.4
It is too tedious to spell out here a non-losing strategy for the O's player, B. Start by dividing A's possible first move into three cases: center, corner, or middle.

Exercise 1.5
The first player A should start by naming 9. Thereafter, when B names x, A should name $10 - x$. Eventually, A reaches the sum of 99. B must then produce a sum that is at least 100.

Exercise 1.6
B has a winning strategy, which consists of doing nothing.

Exercise 1.7
1. A has a winning strategy. A should start by removing two sticks, leaving 12. Thereafter, whenever B removes x sticks, A should remove $4 - x$ sticks. Thus, A always leaves a multiple of four sticks (in fact, 12, 8, 4, 0 sticks).
2. If n is not a multiple of 4, A has a winning strategy. A's first move is to take away as many sticks as is necessary to leave a multiple of four sticks. Then, A operates as in (1). When n is a multiple of 4, B has a winning strategy. A's first move leaves a number of sticks that is not a multiple of 4, and then B can use the winning strategy just indicated.
3. When n is not of the form $4k + 1$, player A has a winning strategy. In such a case, A's first move should leave a number of the form

$4k + 1$. Thereafter, whenever B takes away x sticks, A should then remove $4 - x$ sticks. Eventually A leaves one stick, which B has to remove and lose. When n is of the form $4k + 1$, then player B can use the technique just indicated (that is, whenever A removes x sticks, B should remove $4 - x$ sticks). B eventually leaves one stick, which A has to take and lose.

4. A has a winning strategy. Start by removing four sticks, leaving 10. Thereafter, whenever B removes x sticks, A should then take $5 - x$ sticks. A always leaves a multiple of five sticks (10, 5, 0) and wins.

5. When n is not a multiple of $k + 1$, player A has a winning strategy. A's first move should leave a multiple of $k + 1$ sticks. Thereafter, whenever B takes x sticks, A should take $k + 1 - x$ sticks. Eventually A leaves zero sticks and wins. When n is a multiple of $k + 1$, then A's first move must leave a number that is not a multiple of $k + 1$ and then B can use the winning strategy just described.

6. When n is of the form $j(k + 1) + 1$, B has a winning strategy, namely, whenever A removes x sticks, B should remove $k + 1 - x$ sticks. Eventually, B leaves one stick, which A must take and lose. When n is not of the form $j(k + 1) + 1$, then A has a winning strategy, namely, produce by A's first move a number of the form $j(k + 1) + 1$ and then follow the strategy just described.

Exercise 1.8

1. The first player A has a winning strategy. A should place a penny at the center of the table. Thereafter, whenever B puts a penny down with its center at a certain point P, A should place a penny so that its center is on the point that is symmetric to P with respect to the center of the table. There will always be room for this penny.

2. The second player B has a winning strategy. Whenever A places a domino, B puts down a domino on the squares symmetric to A's domino with respect to the center of the board.

3. Player A has a winning strategy. A first takes the stick at the center of the array. Thereafter, whenever B moves, A then takes away the sticks symmetric with respect to the center to the sticks taken away by B. Since A always can move, B must lose.

4. Player B has a winning strategy, always choosing the group of sticks symmetric (with respect to the center) to the sticks just chosen by A.

Exercise 1.9

1. Both players have non-losing strategies if they imagine playing non-losing strategies for Tic-Tac-Toe on the indicated magic square. Whenever a player places a nought or a cross in a square, that player picks the card whose number is in that square. Since neither player will fill in a horizontal row or vertical column or diagonal, neither player obtains a sum of fifteen.

2. The nine words can be arranged in a square, as in the diagram on the right, where the words in each row, column, or diagonal have exactly one letter in common (and no other triple of words have a letter in common). The two players can play their non-losing strategies in a game of Tic-Tac-Toe on that square and, therefore, they can prevent each other from choosing three words having a letter in common.

PONY	PECK	PUMA
SHIN	EASY	LUST
WARN	BRED	JURY

3. The second player's non-losing strategy is too tedious to verify here.

4. Jill's winning strategy is always obvious, but on a case-to-case basis that is too long to make it worthwhile to describe here.

5. X has a winning strategy. X moves into a corner, and O cannot block the center on the first move. It is easy to show how X can win, no matter what O's first move is.

6. The X's player has a winning strategy. X first should move into the upper left-hand corner. Analysis of the four essentially different first moves by the O's player can show how X forces a win in each case. (One case is pictured in the diagram below. The subscripts indicate the order of the moves.)

X_1	O_3	X_3
O_2	X_5	X_4
X_2	O_4	O_1

7. The first player A has a winning strategy. A should first put a cross in the center square. The second player B is forced to use a nought as a first move. If B puts an O in a corner square, then A should put an O in the opposite corner, and then A will win on A's next turn. If B puts an O in a middle square on B's first move, then A should put an O in the opposite middle square. B is then forced to put an O in another middle square, and A follows by putting an O in the fourth middle square. Whatever B does next, A will win on A's next move.

8. The first player A, the X's player, has a winning strategy, starting with a move into the center square. The O's player B then can move either into a corner or a middle square. The appropriate sequence of moves by A is shown in Diagrams (a) and (b). The subscripts indicate the order of the moves.

(a)

O_1		O_3
O_2	X_1	X_2
X_3		

(b)

O_2	O_1	X_3
	X_1	O_3
		X_2

In both cases, all of B's moves except the first are forced. In (a), A can win by moving X_1 and X_2 down to the bottom row without interference by B and before B can win.

In (b), A can win by moving X_2 into the lower left-hand corner without interference by B and before B can win.

Exercise 1.10

1. Player A has a winning strategy. No matter where B places the penny on the initial edge (say, the bottom row), A can start moving in that row in the direction that will force B to move from an end point on that row into the next row. (The correct direction is the one having an odd number of squares in the row on that side of the square in which A is placed.) Then A has the first move in the next row and can again force B to be the first player to move into the following row. Eventually, B is forced to be the first player to move into the next-to-the-last row. Then A immediately moves into the last row and wins.

2. Player A again has a winning strategy. It is the same as the strategy in (1), except that A continues to force B to be the first player to move into the next row until B is finally forced to move into the last row.

Exercise 1.11

Consider a game with three players, A, B, C. Player A has the first move and that move can go to either of two terminal nodes. Those two nodes have pay-offs $(-1,2,-1)$ and $(-1,-1,2)$, where a triple (a,b,c) indicates pay-offs of a, b, c to players A, B, C, respectively. Player A receives a negative pay-off for either of his two strategies. Players B and C have only the null strategy, that is, they can only sit and watch what A does, and there are possible plays of the game under which they receive negative pay-offs.

Exercise 1.12

A has a winning strategy. Note that the collection $(1,1)$ is a losing position (for the player whose move it is) and that it was just shown that $(2,2)$ also is a losing position. A should first remove two objects from the top pile, producing the collection $(1,2,3)$. Then B has to produce one of the following collections: $(2,3)$, $(1,1,3)$, $(1,3)$, $(1,2,1)$, $(1,2,2)$, and $(1,2)$. A can then produce either $(1,1)$ or $(2,2)$, which are losing positions for B.

Exercise 1.13

$$8 = 2^3, 9 = 2^3 + 2^0, 10 = 2^3 + 2^1, 11 = 2^3 + 2^1 + 2^0, 12 = 2^3 + 2^2,$$

$$14 = 2^3 + 2^2 + 2^1, 15 = 2^3 + 2^2 + 2^1 + 2^0, 16 = 2^4, 17 = 2^4 + 2^0,$$

$$18 = 2^4 + 2^1, 19 = 2^4 + 2^1 + 2^0, 20 = 2^4 + 2^2.$$

Exercise 1.14

1. $3 = 2^1 + 2^0, 5 = 2^2 + 2^0, 3 = 2^1 + 2^0$. So, since there is an odd number of 2^2 terms, the collection is unbalanced and player A has a winning strategy. A's first move should be to remove four objects from the 5 pile.
2. The collection is balanced and, therefore, B has a winning strategy.

Exercise 1.15

Remove five objects from the 6 pile.

Exercise 1.16

When the two piles contain the same number of objects, their binary decompositions will contain exactly the same powers of 2 and so the collection will be balanced and player B has a winning strategy. When the two piles contain different numbers of objects, their binary decompositions must be different. Hence, some power of 2 is in one of the decompositions but not in the other, implying that the collection is unbalanced and player A has a winning strategy.

Exercise 1.17

Going first is preferable. Examination of the triples that add up to 10 shows that the only possible balanced collection is a permutation of 5, 4, 1. So, there are only six collections that give the second player a winning strategy. There are thirty unbalanced collections, all of them giving the first player a winning strategy.

Exercise 1.18

1. Each number from 1 to 15 is listed under the columns headed by the powers of 2 in its binary decomposition. For example, 9 is listed under 8 and 1. If the unknown number is 9, then Peter is told that it is in the first and fourth columns. Hence, the number is $8 + 1 = 9$.
2. The columns would now be headed by 16, 8, 4, 2, and 1.

Exercise 1.19

Existence: Assume every positive integer $< n$ has a binary decomposition. Let j be the largest integer such that $2^j \leq n$. If $n = 2^j$, we are finished. Otherwise, $n - 2^j$ has a binary decomposition and, if we add 2^j to that decomposition, we get a binary decomposition for n.

Uniqueness: Assume positive integers $< n$ have unique decompositions. Assume that we have two decompositions for n. The highest power 2^j in such a decomposition is the highest power of 2 that is $\leq n$ and so it is the same for both decompositions. Subtracting 2^j from both decompositions, we get decompositions for $n - 2^j$, which, by our inductive hypothesis, must be identical. Therefore, the original decompositions for n were identical.

Exercise 1.20

1. Player A has a winning strategy. A first removes the 5 pile.
2. Player A's winning strategy starts by removing the 6 pile.
3. Player A's winning strategy begins by removing the 2 pile.
4. B has the winning strategy.

Exercise 1.21

B can force a win except in one case. If the original collection is balanced, B can play his usual winning strategy and can tell A to make any move. If the original collection is unbalanced, then an analysis of the various possibilities shows that, with one exception, B can instruct A how to produce another unbalanced collection, for which B has a winning strategy. The exception occurs when the initial collection consists of an odd number of unary piles. In that case, all moves are forced and A comes out the winner.

Exercise 1.38

There are non-losing strategies for both players.

Exercise 1.39

There is a winning strategy for B.

Exercise 1.40

There is a winning strategy for A.

Exercise 1.43

1. (a) $g(3,4) = 2$. Remove two objects from each pile, since $g(1,2) = 0$.
   (b) $g(3,5) = 0$. This is a losing position.
   (c) $g(4,7) = 0$. This is a losing position.
   (d) $g(5,7) = 1$. Remove one object from the 5-pile, since $g(4,7) = 0$.

2.

0	1	2	3	4	5	6	7	8	9
1								6	10
2								7	11
3								10	7
4								1	8
5								2	12
6								0	2
7								3	13
8	6	7	10	1	2	0	3	4	5
9	10	11	7	8	12	2	13	5	0

Exercise 1.44

1. Since $b_9 = 0$, this is a losing position.
2. $b_3 \oplus b_4 \oplus b_8 = 1 \oplus 2 \oplus 1 = 2$. Remove four objects from the 4-pile.
3. $b_2 \oplus b_3 \oplus b_1 \oplus b_2 = 1 \oplus 1 \oplus 0 \oplus 1 = 1$. Remove the 3-pile.

Exercise 1.45

1. $c_4 \oplus c_6 \oplus c_8 = 0 \oplus 1 \oplus 2 = 3$. Change the 8-pile to one with Grundy number 1, for example a 6-pile.
2. $c_4 \oplus c_5 \oplus c_8 = 0 \oplus 2 \oplus 2 = 0$. This is a losing position.
3. $c_3 \oplus c_5 \oplus c_5 \oplus c_8 = 1 \oplus 2 \oplus 2 \oplus 2 = 3$. Remove two objects from the 8-pile.

Exercise 1.46

$c_{11} = 2, c_{12} = 1, c_{13} = 3, c_{14} = 2, c_{15} = 1, c_{16} = 3, c_{17} = 2, c_{18} = 4, c_{19} = 3, c_{20} = 0.$

Hence

$$c_{13} \oplus c_{14} \oplus c_{18} = 3 \oplus 2 \oplus 4 = 5.$$

Reduce the 18-pile by removing three objects, since $c_{15} = 1$.

Exercise 1.47

1. $g(0) = 0$. If $n > 0$, $g(n) = 1$ when n is even and $g(n) = 0$ when n is odd.
2. If the piles of a collection p contain $n_1, n_2, ..., n_k$ objects, respectively, then $g(p) = g(n_1) \oplus g(n_2) \oplus ... \oplus g(n_k)$. p is a winning position if and only if $g(p) > 0$, which holds if and only if the number of 1's in the Nim-sum is odd, which, in turn, holds if and only if the number of even piles is odd.

Exercise 1.48

$g(n)$ is the remainder upon division of n by 3.

Exercise 1.49

The Nim collection is 2 4 4 5 8. This is unbalanced. We must change the 8-pile to a 7-pile by moving the chip in the 8-cell one cell to the left: 2 4 4 5 7. This is balanced.

Exercise 1.50

The gaps are 3 4 1 0 4 1 1 1. This is unbalanced. Remove the 3-pile by moving the white chip in the top row three cells to the right. This yields 0 4 1 0 4 1 1 1, which is balanced.

Exercise 1.51

The gaps are 3 1 0 1 0 1 2. The alternate gaps are 3 0 0 2. This is unbalanced. Reduce the 3-pile to a 2-pile by moving the rightmost chip one cell to the left. So move the black token in the bottom row one cell to the left.

Exercise 1.52

$$8 \oplus 9 \oplus 12 = 13$$

Change the 11-pile to a 1-pile by removing 10 objects. This yields a collection with Grundy number 0.

Exercise 2.1

(a)

	1	2
1	2	-2
2	-1	4

(b)

	H	T
H	0	0
T	0	0

(c)

	1	2
1	2	3
2	3	-4
3	-4	5

Exercise 2.2

Let c_{ij} be the entry in the ith row, jth column. (a) In Example 2.4, the first matrix has saddle points c_{11}, c_{13}; the second matrix has saddle points at the four 1's in the first and third rows. (b) c_{21} (c) None (d) c_{22} (e) c_{12}, c_{14}, c_{22}, c_{24}

Exercise 2.3

Look at Exercise 2.2(b).

Exercise 2.4

Let $c = 30{,}000$. The saddle points are $(c - 1, c - 1)$, $(c - 1, c)$, $(c, c - 1)$, (c, c).

Exercise 2.5

(a) $0 \le x \le 1$ and $v = x$, or $x < 0$ and $v = 0$

(b) $x \ge \frac{1}{3}$, $v = 1$ (c) $x \le \frac{1}{2}$, $v = 1$

Exercise 2.6

(a) maximin $= 4$, minimax $= 4$, saddle point c_{31}, $v = 4$

Exercise 2.7

3. (i) $\begin{pmatrix} 3 & -1 & 0 \\ 0 & 3 & 1 \\ 2 & 0 & 1 \end{pmatrix}$

(ii) (2), saddle point $c_{21} = 2$ (iii) $\begin{pmatrix} 1 & 3 \\ 2 & 1 \end{pmatrix}$ (iv) not reducible, saddle point $c_{11} = 1$

Exercise 2.8

	T	U	V
T	0	$-c/2$	$c/2$
U	$c/2$	0	$c/2$
V	$-c/2$	$-c/2$	0

Both should build at U.

Exercise 2.10
   2. $cd \le 0$

Exercise 2.11
   1. 19/12

Exercise 2.18
(a) $X = (2/5, 2/5, 1/5) = Y$. (b) $X = (1/2, 1/8, 3/8) = Y$, (d) $X = (1/4, 0, {}^1\!/_2, {}^1\!/_4)$

Exercise 2.20
   1. $X = Y = (2/7, 4/7, 1/7)$

Exercise 2.21
(a) $Y = ({}^1\!/_2, 0, {}^1\!/_2)$, $v = 2$, $X = tX_1 + (1 - t)X_2$ for $0 \le t \le 1$, where $X_1 = (1/2, 0, 1/2)$ and $X_2 = (1/3, 1/3, 1/3)$
(b) $Y = (2/3, 0, 1/3)$, $v = 1$, $X = tX_1 + (1-t)X_2$ for $0 \le t \le 1$, where $X_1 = (1/3, 1/3, 1/3)$ and $X_2 = 1/2, 1/2, 0)$
(c) $X = Y = (bc/K, ac/K, ab/K)$ with $K = bc + ac + ab$
(d) $X = (1/2, 3/8, 1/8)$, $Y = (3/17, 2/17, 12/17)$, $v = 2$
(e) $X = (2/3, 1/3, 0)$, $Y = (5/6, 1/6, 0)$, $v = 1/3$.

Exercise 2.22

1. $\begin{pmatrix} 1 & -1 & -1 \\ -2 & 2 & -2 \\ -3 & -3 & 3 \end{pmatrix}$, $X = (6/11, 3/11, 2/11)$, $Y = (5/22, 4/11, 9/22)$, $v = -6/11$

2. $\begin{pmatrix} 1 & -1 & -2 \\ -1 & 2 & -1 \\ -2 & -1 & 3 \end{pmatrix}$, $X = Y = (15/31,7/31,9/31)$, $v = -10/31$

Exercise 2.26
(a) $X = (2/7,5/7)$, $Y = (6/7,1/7)$, $v = 635/7$
(b) $X = (3/29,26/29)$, $Y = (24/29,5/29)$, $v = 12/29$
(c) $X = (2/7,0,5/7)$, $Y = (6/7,0,1/7)$, $v = 23/7$
(d) $X = (0,3/4,1/4)$, $Y = (1/2,1/2,0)$, $v = {}^1\!/_2$
(e) $X = (0,1)$, $Y = (1,0)$, $v = 6$
(f) $X = (0,1/4,3/4)$, $Y = (1/4,0,3/4)$, $v = 5/4$.

Exercise 2.27
1. $\begin{pmatrix} -4 & K \\ K & -9 \end{pmatrix}$
2. None
3. $K = 6$
4. Melanie: $(3/5,2/5)$, Peter: $(3/5,2/5)$

Exercise 2.28
1.

	H	T
H	4	-10
T	-5	9

2. $X = (1/2,1/2)$, $Y = (19/28,9/28)$
3. $v = -1/2$, Not fair

Exercise 2.29
2. If B plays $(1/6,1/3,1/2)$, the best strategy for A is $(0,0,1)$. If A plays $(1/2,0,1/2)$, the best strategy for B is $(0,0,1)$.

Exercise 2.30
1. $X = (5/7,0,2/7)$
2. A_3

Exercise 2.31
(a) $X = (1,0)$, Y arbitrary

(b) $X = (1,0)$, $Y = (1,0)$

(c) $X = (1,0)$, $Y = (1,0)$

Exercise 2.33

$$\begin{pmatrix} 1 & 2 & 1 \\ 0 & 1 & 2 \\ 0 & 1 & 2 \end{pmatrix}$$

has a saddle point, but its transpose doesn't.

Exercise 2.34
1. $X = (1/4,3/4)$, $Y = (5/8,3/8)$, $v = 7/4$
2. (i) 80 (ii) 140 (iii) 240 (iv) 140

Exercise 2.35
(a) $X = (1/2,1/2)$, $Y = (1/4,3/4,0)$
(b) $X = (2/5,3/5)$, $Y = (1/5,4/5,0,0)$
(c) $X = (9/14,5/14)$, $Y = (1/14,13/14,0,0)$

Exercise 2.36
(a) $X = (5/7,2/7)$, $Y = (0,6/7,1/7)$
(b) $X = (x,1-x)$ for $1/3 \le x \le 1/2$, $Y = (1,0,0)$
(c) $X = (x,1-x)$ for $1/2 \le x \le 2/3$, $Y = (0,1,0,0)$

Exercise 2.37
(a) $X = (0,1,0)$, $Y = (y,1-y)$ for $1/3 \le y \le 2/3$, $v = 1$
(b) $X = (2/5,3/5,0)$, $Y = (4/5,1/5)$, $v = 8/5$
(c) $X = (2/5,3/5,0)$, $Y = (4/5,1/5)$, $v = 12/5$

Exercise 2.38
The company should follow its normal procedure 7/13 of the time and take the insurance contract 6/13 of the time. Its expected cost per horn would be $(56/13).

Exercise 2.39
(a) $X = (2/3,1/3,0)$, $Y = (1/3,0,2/3)$, $v = 7/3$
(b) Saddle point $X = (0,1,0)$, $Y = (1,0,0)$, $v = 2$
(c) $Y = (0,1/2,1/2)$, $v = 1$, $X = (0,1,0) + u(1,-2,1)$ for $0 \le u \le 1/3$
(d) $X = (1/2,1/2,0)$, $v = 2$, $Y = (0,1,0) + u(1,-2,1)$ for $1/3 \le u \le 1/2$
(e) $X = (5/13,6/13,2/13)$, $Y = (1/13,7/13,5/13)$, $v = 35/13$

Exercise 2.41
1. Maximum 15 at (3,0), minimum −6 at (0,2)
2. Maximum 12 at (2,0), minimum 0 at (0,0)
3. Maximum − 13/7 at (9/7,8/7), minimum −25 at (0,5)

Exercise 2.42

1. No maximum
2. Minimum 1 along line segment joining (1,1) and (1,0)
3. Maximum 15 at (5,0)
4. Minimum 6 at (2,0)
5. No maximum
6. Empty constraint set

Exercise 2.43

1. Minimize $v = 2a - 4b + 5c + 4d$ subject to the constraints $a + 3b + 2c + d \geq 3$, $-a + 3b - 3c + d \geq 9$, $a - 2b - c + 2d \geq -1$
2. Maximize $u = 7x$ subject to the constraints $2x - 4y \leq 3$, $5x - y \leq -4$
3. Minimize $v = 3a - 2b$ subject to the constraints $3a + 7b \geq 2$, $a - 4b \geq -1$, $-2a \geq -1$, $-a + b \geq 4$

Exercise 3.2

(1), (4), and (5)

Exercise 3.3

Maximize $u = 12x - 4y$, subject to $3x - 4y - 2 = -r$ and $2x + 3y - 4 = -s$

Exercise 3.4

The same one

Exercise 3.5

1. No solution
2. Maximum $= 9/2$ at $x = 0$, $y = 3/2$
3. Maximum $= 2$ at $x = 2$, $y = z = 0$
4. No solution
5. Maximum $= 8$ at $x = 1$, $y = z = w = 0$
6. Maximum $= 4$ at $x = 0$, $y = 2$, $z = 9$
7. Maximum $= 11$ at $x = 3$, $y = 0$, $z = 0$
8. Maximum $= 10$ at $x = 6/7$ and $y = 20/7$
9. Maximum $= 188/9$ at $x = 0$, $y = 22/9$, $z = 17/9$

Exercise 3.7

1. Maximum $= 7$ at $x = y = z = 0$ or $x = 0$, $y = 1$, $z = 0$ and all points in the included segment.
2. Maximum $= 9$ at $x = 0$, $y = 3$ and $x = 1/4$, $y = 21/8$, and all points in the included segment.

Exercise 3.8

1. No solution
2. Maximum $= 9/2$ at $x = y = 0$, $z = 3/2$
3. No solution
4. Minimum $= 191/7$ at $x = 25/7$, $y = 24/7$, $z = 0$

Exercise 3.9
(a) $a_{22} = 3/2$
(b) $a_{32} = 3$

Exercise 3.10
    1.   Optimal value = 76/11 at $x_1 = 17/11$, $x_2 = 2/11$, $y_1 = 5/11$, $y_2 = 23/11$
    2.   No solution

Exercise 3.11
(a) The equilibrium pair (X, Y) is $X = (1/5, 2/5, 2/5)$, $Y = (0, 2/5, 2/5, 1/5)$, and the value v is $1/5$.
(b) The equilibrium pair (X, Y) is $X = (3/7, 4/7)$, $Y = (0, 2/7, 0, 5/7)$, and the value v is $6/7$. (This also could be solved by our earlier graphing technique for $2 \times n$ matrices.)

Exercise 3.12
(a) $X = (4/9, 0, 1/9, 0, 4/9)$, $v = 14/9$, and $Y = (7/90, 32/90, 48/90, 3/90)$. Although X is symmetric with respect to equivalent strategies $(0,4)$ and $(4,0)$, and $(1,3)$ and $(3,1)$, this symmetry is lacking in Y with respect to equivalent strategies $(0,3)$ and $(3,0)$, and $(1,2)$ and $(2,1)$. However, the strategy $Y^* = (3/90, 48/90, 32/90, 7/90)$, obtained by switching $(0,3)$ and $(3,0)$, and $(1,2)$ and $(2,1)$, is also optimal. Then all strategies on the segment between Y and Y^* also are optimal. The midpoint of that segment is the symmetric optimal strategy $Y^{**} = (1/18, 8/18, 8/18, 1/18)$.
(b) $X = (7/15, 0, 1/5, 1/3)$, $Y = (1/5, 3/5, 1/5)$, $v = 6/5$. Note that $X^* = (1/3, 1/5, 0, 7/15)$ is also optimal. So, all the vectors on the segment between X and X^* are optimal. In particular, the midpoint of that segment $(2/5, 1/10, 1/10, 2/5)$ is optimal and shows the symmetry between strategies $(0,3)$ and $(3,0)$, and between $(1,2)$ and $(2,1)$.

Exercise 4.1
Yes for Examples 4.1 and 4.2.

Exercise 4.2
(a) A's maximin is 1 at strategy A_2. B's maximin is 3 at B_3. The maximin pair is (A_2, B_3). It is not stable. (b) A's maximin is 0 at A_1. B's maximin is 2 at B_1. The maximin pair is (A_1, B_1). It is not stable.

Exercise 4.4
In Example 4.1, (A_2, B_1) is the unique Nash equilibrium, there is no among-the-best strategy pair, and (A_1, B_1), (A_3, B_1), and (A_3, B_2) are Pareto optimal. In Example 4.2, (A_1, B_2) is among-the-best and, therefore, a Nash equilibrium and Pareto optimal.

Exercise 4.5
Example 4.1.

Exercise 4.9
In Example 4.3, (A_1,B_2) and (A_2,B_1). In Exercise 4.2(c), (A_1,B_2) and (A_2,B_3).

Exercise 4.10
(a) $X = (2/5,3/5)$, $Y = (1/3,2/3)$, pay-offs $(5/3,8/5)$
(b) $X = (2/3,1/3)$, $Y = (1/2,1/2)$, pay-offs $(7/2,8/3)$
(f) Pure Nash equilibrium (A_1,B_1) with pay-offs $(2,4)$. No mixed Nash equilibrium.
(g) Pure Nash equilibria: (A_1,B_1) with pay-offs $(2,4)$ and (A_2,B_2) with pay-offs $(4,1)$. Mixed Nash equilibrium: $X = (1/2,1/2)$, $Y = (3/4,1/4)$ with pay-offs $(7/4,2)$.

Exercise 4.11
(a) Pure Nash equilibria: (A_1,B_1) with pay-offs $(3,3)$ and (A_2,B_2) with pay-offs $(0,0)$. Mixed Nash equilibrium: $X = (1/4,3/4)$, $Y = (1/4,3/4)$, pay-offs $(0,0)$.
(b) No pure Nash equilibrium. Mixed Nash equilibrium: $X = (1/4,3/4)$, $Y = (1/2,1/2)$ with pay-offs $(0,0)$.
(c) Pure: (A_1,B_2) with pay-offs $(1,3)$, and (A_2,B_1) with pay-offs $(1,4)$. Mixed: $X = (4/5, 1/5)$, Y : arbitrary, pay-offs $(1,12/5)$.
(e) Mixed: $X = (x,0,0,1 - x)$ for $3/4 \le x \le 1$, $Y = (0,1,0)$, pay-offs $(3,2x + 1)$
(g) Mixed: $X = (3/4,1/4,0)$. $Y = (0,1/2,1/2)$, pay-offs $(3/2,3/2)$.

Exercise 4.14
(a) Mixed: $X = (1/3,1/3,1/3)$, $Y = (2/14,1/14,11/14)$, pay-offs $(23/14,2)$ (b) Mixed: $X = (5/16,1/4,7/16)$, $Y = (1/4,3/4,0)$, pay-offs $(9/4,3/4)$
(d) Pure: (A_1,B_2) with pay-offs $(3,4)$; (A_2,B_3) with pay-offs $(4,2)$. Mixed: $X = (0,1,0)$, $Y = (0,3/4,1/4)$ with pay-offs $(7/4,2)$
(e) Mixed: $X = (1/2,1/2,0)$, $Y = (a,^1/_2 - a,^1/_2)$ with $a \le ^1/_2$; pay-offs $((9/2) - 2a, 3/2)$
(f) Pure Nash equilibria at (A_1,B_1), (A_2,B_2), (A_3, B_3), (A_4,B_4) with respective pay-offs $(1,4)$, $(2,3)$, $(3,2)$, $(4,1)$. Mixed Nash equilibrium $X = (3/25,4/25,6/25,12/25)$, $Y = (12/25,6/25,4/25,3/25)$ with pay-offs $(12/25,12/25)$.

Exercise 4.15
(a) Mixed: $X = (1/3,1/3,1/3)$, $Y = (1/3,1/3,1/3)$ with pay-offs $(1,1)$ (d) Mixed: $X = (1/2,1/2)$, $Y = (2/5,b,(3/5) - b)$ with $0 \le b \le 3/5$; pay-offs $((9/5) - b,3/2)$.
(f) Mixed: $X = (4/35,20/35,11/35)$, $Y = (1/2,0,1/2)$, pay-offs $(7/2,99/35)$.
(i) Pure: (A_2,B_1) with pay-offs $(4,5)$. Mixed: $X = (3/4,1/4)$, $Y = (0,1/7,6/7)$, pay-offs $(11/7,9/4)$.

Exercise 4.16
(e) Mixed: Nash equilibrium: $X = (1000/1001,1/1001)$, $Y = (1/2,1/2)$, with pay-offs $(5/2,1000/1001)$. Player A can threaten to play strategy A_2, with dire consequences for B, but the optimal bribe is not clearly determined.

Exercise 4.21
1. For the game in Exercise 4.2(b), there are no pure Nash equilibria. There is a mixed Nash equilibrium $X = (2/3,1/3)$, $Y = (1/3,2/3)$ with

pay-offs (1/3,7/3). The set $\mathcal{R}$ of pay-off points consists of all pairs (u,v) such that $u = 3xy - x - 2y + 1$, $v = -3xy + 2x + 2y + 1$ for $0 \le x \le 1$, $0 \le y \le 1$. The set $\mathcal{R}^*$ of pay-offs for joint randomized strategies is the boundary and interior of the quadrilateral $RSTU$, where $R = (-1,3)$, $S = (0,3)$, $T = (1,2)$, $U = (1,1)$. The negotiation set is the line segment ST.

3. For the game in Exercise 4.10(a), there is no pure Nash equilibrium. There is a mixed Nash equilibrium $X = (2/5,3/5)$, $Y = (1/3,2/3)$, with pay-offs (5/3,8/5). The set $\mathcal{R}$ of pay-off points consists of all pairs (u,v) such that $u = -3xy + x + 2y + 1$, $v = 5xy - x - 2y + 2$ for $0 \le x \le 1$, $0 \le y \le 1$. The set $\mathcal{R}^*$ of pay-offs for joint randomized strategies is the boundary and interior of the triangle RST, where $R = (1,4)$, $S = (3,0)$, $T = (1,2)$. The negotiation set is the segment RS. For the game in Exercise 4.10(f), there is a pure Nash equilibrium (A_1,B_1) with pay-offs (2,4), but there is no mixed Nash equilibrium. The set $\mathcal{R}$ of pay-off points consists of all pairs (u,v) such that $u = -3xy + 4x + y$, $v = 3x + y + 1$ for $0 \le x \le 1$, $0 \le y \le 1$. The set $\mathcal{R}^*$ of pay-offs for joint randomized strategies is the boundary and interior of the triangle RST, where $R = (2,4)$, $S = (4,3)$, $T = (0,1)$ The negotiation set is the segment RS.

Exercise 4.22

(b) There is no pure Nash equilibrium. There is one mixed Nash equilibrium $X = (1/2,1/2)$, $Y = (9/14,5/14)$ with pay-offs (11/2,7/2) The set $\mathcal{R}$ of pay-off points consists of all pairs (u,v) such that $u = 14xy - 9x - 7y + 10$, $v = -2xy + 3x + y + 2$ for $0 \le x \le 1$, $0 \le y \le 1$. The set $\mathcal{R}^*$ of pay-offs for joint randomized strategies is the boundary and interior of the quadrilateral $RSTU$, where $R = (1,5)$, $S = (8,4)$, $T = (10,2)$, $U = (3,3)$. The negotiation set is the broken line RST. Players A and B can choose joint randomized strategies that will maximize the sum of their pay-offs by selecting a point on ST. The sum will be 12, but we have not established a procedure for choosing a "fair" division of that sum. Although A's pay-off on ST varies between 8 and 10, B can argue against giving A as much as 8, since there are points in the negotiation set on segment RS that will yield A as little as 1. Note that, as in earlier cases, there is no obvious reason why the pay-offs (11/2,7/2) at the mixed Nash equilibrium should play a special role in this decision process.

Exercise 4.25

For Example 4.21, the matrix for A is

$$\begin{pmatrix} 4 & 2 \\ 1 & 5 \end{pmatrix}$$

so that A's support level is $18/6 = 3$. (18 is the value of the determinant, and $6 = (4 + 5) - (1 + 2)$.) The matrix for B is

$$\begin{pmatrix} 3 & 1 \\ 5 & 0 \end{pmatrix}$$

so that B's support level is $-5/-3 = 5/3$.

For Exercise 4.21(6), the matrix for A is

$$\begin{pmatrix} 8 & 1 \\ 3 & 10 \end{pmatrix}$$

so that A's support level is $77/14$.

The matrix for B is

$$\begin{pmatrix} 4 & 3 \\ 5 & 2 \end{pmatrix}$$

This has a saddle point with value 3, so that B's support level is 3.

Exercise 4.26
    2. The negotiation set consists of a single point $(3,5)$, which is the chosen point.
    3. SQ is $(13/5,5/2)$. The arbitration procedure yields the pay-offs $(71/20,69/20)$.

Exercise 4.27
    1. The arbitration procedure yields pay-offs $(29/28,81/28)$.

Exercise 4.28
    6. (i) $(3,3)$ (ii) $(13/3,7/3)$

Exercise 4.30
(a) (A_2,B_2,C_1) with pay-offs $(2,2,2)$ (b) (A_1,B_2,C_1) with pay-offs $(0,1,1)$, and (A_2,B_1,C_2) with pay-offs $(0,0,1)$

Exercise 4.31
No pure Nash equilibrium. Mixed Nash equilibrium: $X = (2/3,1/3)$, $Y = (3/4,1/4)$, $Z = (4/15,11/15)$ with pay-offs $(-3/20,2/5,13/12)$. The coalitions have the following values: $v(AB) = 1/2$, $v(BC) = 2/5$, $v(AC) = 1/3$, $v(ABC) = 0$. A prefers a coalition with B, B prefers a coalition with A, and C prefers a coalition with B.

Exercise 4.32

(a) Pure Nash equilibria (A_1, B_1, C_2) with pay-offs $(0,1,0)$, (A_1, B_2, C_1) with pay-offs $(0,0,1)$, and (A_2, B_1, C_1) with pay-offs $(1,0,1)$. There is more than one mixed Nash equilibrium, e.g., $X = (1/2, 1/2)$, $Y = (0,1)$, $Z = (0,1)$ with pay-offs $(1/2, 1/2, 0)$; $X = (1/2, 1/2)$, $Y = (1,0)$, $Z = (1,0)$ with pay-offs $(0,0,1/2)$. Using support levels, $v(A) = 0$ and $v(BC) = \frac{1}{2}$; $v(B) = 0$ and $v(AC) = 2$; $v(C) = -\frac{1}{2}$ and $v(AB) = \frac{1}{2}$; $v(ABC) = 2$. A and C prefer a coalition with each other, B prefers a coalition with A or C to playing alone.

Exercise 4.33

For the game of Exercise 4.32(b)

$$v(A) = 0, \; v(BC) = 1, \; v(B) = 0$$

$$v(AC) = 1, \; v(C) = 2/3, \; v(AB) = 7/6, \; v(ABC) = 1$$

Exercise 4.36

2. The seller has Shapley value $k/(k+1)$ and each buyer has Shapley value $\dfrac{1}{k(k+1)}$.

Exercise 4.37

1. The Vice-President and each senator has Shapley value $1/101$.
2. The chairperson and each member has Shapley value $1/(2k+1)$.

Exercise 4.38

2. The chairperson has Shapley value $1/k$ and each of the other members has Shapley value $\dfrac{k-1}{k(2k-1)}$.

Exercise 4.39

1. The Shapley value of a_3 is 0 and each of a_1 and a_2 has Shapley value $\frac{1}{2}$.
2. The Shapley value of a_3 is 0 and each of a_1 and a_2 has Shapley value $5/2$.
3. (i) A, B, C have Shapley values $2/3$, $1/6$, $1/6$, respectively. (ii) A, B, C have Shapley values $1/3$, $1/3$, $1/3$, respectively. (iii) A, B, C, D have Shapley values $5/12$, $3/12$, $3/12$, $1/12$, respectively.

Exercise 4.42

1. The core is the set of all imputations, that is, the set of all pairs (x_1, x_2) of nonnegative real numbers such that $x_1 + x_2 = 1$.
2. The core is the set of all imputations, that is, the set of all triples (x_1, x_2, x_3) of nonnegative real numbers such that $x_1 + x_2 + x_3 = 1$.

3. If N contains n members, then the core contains just one element, the n-tuple $(1,1,\ldots, 1)$ all of whose coordinates are 1.

Exercise 4.43

2. It suffices to show $v(S) = \sum_{a_j \in S} v_j$ for every coalition S, and this can be done through a proof by contradiction. In fact, $v(S) \geq \sum_{a_j \in S} v_j$ and $v(N - S) \geq \sum_{a_j \in N-S} v_j$ by superadditivity. So, if $v(S) = \sum_{a_j \in S} v_j$ is false, then $v(S) > \sum_{a_j \in S} v_j$, whence $v(S) + v(N - S) > \sum_{j=1}^n v_j = v(N)$, contradicting superadditivity.

3. By part (2), we must have $v(N) > \sum_{j=1}^n v_j$. (Superadditivity yields $v(N) \geq \sum_{j=1}^n v_j$.) Let $\delta = v(N) - \sum_{j=1}^n v_j > 0$. Choose any n-tuple $(y_1,\ldots, y_n)$ of nonnegative real numbers such that $\sum_{j=1}^n y_j = \delta$. (There are infinitely many such n-tuples.) Let $x_j = v_j + y_j$ for $1 \leq j \leq n$. Then $X = (x_1,\ldots, x_n)$ is an imputation.

Exercise 4.44

1. Assume $(x_1,\ldots,x_n)$ is an imputation. Then $v(N) = \sum_{j=1}^n x_j \geq \sum_{j=1}^n v_j = v(N)$. So, $x_j = v_j$ for $1 \leq j \leq n$. Thus, $(v_1,\ldots, v_n)$ is the only imputation and is in the core (since $v(S) = \sum_{a_j \in S} v_j$). Hence, the set of imputations is identical with the core and consists of just one n-tuple $(v_1,v_2,\ldots, v_n)$.

2. Since a Shapley value is an imputation, part (1) implies that the Shapley value of an inessential game is $(v_1,\ldots, v_n)$.

Exercise 4.46

(a) Essential. $v_1 = v(A) = {}^1\!/_2$, $v_2 = v(B) = 1$, $v(N) = 5 \neq v_1 + v_2$ (b) Inessential. $v_1 = v(A) = 3/2$, $\mathrm{v}_2 = v(B) = -{}^1\!/_2$, $v(N) = 1 = v_1 + v_2$. Constant-sum (c) Essential. $v_1 = v(A) = 0$, $v_2 = v(B) = 1$, $v_3 = v(C) = 1/3$, $v(N) = 3 \neq v_1 + v_2 + v_3$ (d) Essential. $v_1 = v(A) = 1/3$, $v_2 = v(B) = -6/7$, $v_3 = v(C) = -7/5$, $v(N) = 0$. Zero-sum(e) Inessential. $v_1 = v(A) = -{}^1\!/_2$, $v_2 = v(B) = 5/2$, $v(N) = 2 = v_1 + v_2$. Constant-sum.

Exercise 4.47

Let $N = \{a_1,a_2\}$. $v(N) = 0$, since the pay-offs for $\{a_1,a_2\}$ are zero. But $v_1 = -v_2$ by von Neumann's Theorem for two-person zero-sum games. So, $v(N) = v_1 + v_2$.

Exercise 4.48

3.

$$\begin{pmatrix} (0,0) & (0,-1) \\ (0,0) & (0,-1) \end{pmatrix}$$

As a matrix game, this is not zero-sum. However, the corresponding coalition game is zero-sum:

$$v_1 = v(A) = 0, \; v_2 = v(B) = 0, \; v(N) = 0$$

Exercise 4.50

4. (ii) Let d be the reciprocal of $v(N) - \sum_{j=1}^{n} v_j$ and let $e_j = -d\, v_j$ for $1 \le j \le n$. Define the coalition game (N,u) by setting $u(S) = d(v(S)) + \sum_{a_j \in S} e_j$ for all coalitions S. Then (N,u) is strategically equivalent to (N,v), and (N,u) is $\{0,1\}$-normalized.

5. The game of Example 4.29 is strategically equivalent to the $\{0,1\}$-normalized game (N,u), with $N = \{a_1,a_2,a_3\} = \{A,B,C\}$, where $u(\{B,C\}) = u(\{A,C\}) = u(\{A,B\}) = 1$ This is just the three-person majority game. The game of Example 4.30 is strategically equivalent to the $\{0,1\}$-normalized game (N,w), where $w(\{B,C\}) = 1/5$, $w(\{A,C\}) = 1/15$, and $w(\{A,B\}) = 3/5$.

Exercise 4.53

1. Consider the four-person coalition game (N,v) in which $N = \{a_1,a_1,a_3,a_4\}$, $v(S) = 1$ for all coalitions S containing at least two members, and $v(S) = 0$ otherwise. Let $X = (1/2,1/2,0,0)$ and $Y = (0,0,1/2,1/2)$. Then X dominates Y over $\{a_1,a_2\}$ and Y dominates X over $\{a_3,a_4\}$.

2. In the game of part (1), X dominates Y and Y dominates X, but X does not dominate X.

Exercise 4.57(b)

We know that the core is a subset of *UND*. For the converse, assume $X = (x_1,\dots, x_n)$ is an imputation that is not in the core. We must show that it is not in *UND*. Since X is not in the core, there is some coalition S for which $v(S) > \sum_{a_j \in S} x_j$. Clearly S is nonempty.

Let $\varepsilon = v(S) - \sum_{a_j \in S} x_j$. Now define $Y = (y_1,\dots, y_n)$ as follows: For j in S, let $y_j = x_j + \dfrac{\varepsilon}{|S|}$ and, for j not in S, let $y_j = v_j + \left(\dfrac{1}{n-|S|}\right)(v(N) - v(S) - \sum_{a_j \in S} v_j)$. Then Y is an imputation and Y dominates X over S. Hence, X is not in *UND*.

Exercise 4.58

Consider the three-person coalition game (N,v) in which $v(N) = 4$, $v(S) = 1$ for every two-person coalition S, and $v(S) = 0$ for every one-person coalition S.

Bibliography

This list contains not only books and papers mentioned in the text, but also other items that the reader may find of interest.

Adelson-Velsky, G.M., V.L. Arlazarov, and M.V. Donskoy, *Algorithms for Games*, Springer, 1988.

Allen, L.E., Games bargaining: a proposed application of the theory of games to collective bargaining, *Yale Law Journal*, Vol. 65, 1956, pp. 660–693.

Alpern, S. and A. Beck Hex games and twist maps on the annulus, *American Mathematical Monthly*, Vol. 98, 1991, pp. 803–813.

Ankeny, N.C., *Poker Strategy: Winning with Game Theory*, Basic Books, 1981.

Arrow, K.J., *Social Choice and Individual Values*, Wiley, 1951.

Aumann, R.J., A survey of cooperative games without side payments, in *Essays in Mathematical Economics*, M. Shubik (Ed.), Princeton University Press, 1967, pp. 3–27.

Aumann, R.J., Game theory, in *The New Palgrave: A Dictionary of Economics*, Macmillan, 1987, pp. 460–482.

Aumann, R. J., *Lectures on Game Theory*, Westview Press, 1989.

Aumann, R.J., Nash equilibria are not self-enforcing, in *Economic Decision-Making: Games, Econometrics and Optimisation*, Gabszewicz, J.J., J.-F. Richard, and L.A. Wolsey (Eds.), Elsevier, 1990, pp. 201–206.

Aumann, R.J. and J. Drèze, Cooperative games with coalition structures, *International Journal of Game Theory*, Vol. 4, 1975, pp. 217–237.

Aumann, R.J. and M. Maschler, The bargaining set for cooperative games, in *Advances in Game Theory*, Vol. 52, *Annals of Mathematics Studies*, Princeton University Press, 1964, pp. 443–476.

Averbach, B. and D. Chein, *Mathematics: Problem Solving Through Recreational Mathematics*, Freeman, 1980.

Axelrod, R., *The Evolution of Cooperation*, Basic Books, 1984.

Axelrod, R., *The Complexity of Cooperation*, Princeton University Press, 1997.

Axelrod, R. and D. Dion, The further evolution of cooperation, *Science*, Vol. 242, 1988, pp. 1385–1390.

Bacharach, M., *Economics and the Theory of Games*, Westview Press, 1977.

Baird, D., R. Gertner, and R. Picker, *Game Theory and the Law*, Harvard University Press, 1994.

Balinski, M.L. and H.P. Young, *Fair Representation*, Yale University Press, 1982.

Ball, W.W.R. and H.S.M. Coxeter, *Mathematical Recreations and Essays*, Thirteenth Edition, Dover, 1987.

Banzhaf, J.F. III, Weighted voting does not work: a mathematical analysis, *Rutgers Law Review*, Vol.19, 1965, pp. 317–343.

Barbera, S., P. J. Hammond, and C. Seidl (Eds.), *Handbook of Utility Theory*, Kluwer, 1999.

Beasley, J.D., *The Mathematics of Games*, Oxford University Press, 1989.

Beck, J., *Combinatorial Games*, Cambridge University Press. 2003.

Beck, A., M. Bleicher, and D. Crowe, *Excursions into Mathematics*, Worth, 1969. (Republished by A.K. Peters, 2000)

Benoit, J.-P. and L.A. Kornhauser, Game-theoretic analysis of legal rules and institutions, in *Handbook of Game Theory with Economic Applications*, Vol. 3, Elsevier, 2002, pp. 2229–2269.

Berlekamp, E.R., J.H. Conway, and R.K. Guy, *Winning Ways for Your Mathematical Plays: Games in General*, Vol. 1, Academic, 1982. (2nd ed., A.K. Peters, 2001)

Berlekamp, E.R., J.H. Conway, and R.K. Guy, *Winning Ways for Your Mathematical Plays: Games in Particular*, Vol. 2, Academic, 1982. (2nd ed., A.K. Peters, 2003)

Binmore, K., *Fun and Games: A Text in Game Theory*, D.C. Heath, 1992.

Bishop, R.L., Game theoretic analysis of bargaining, *Quarterly Journal of Economics*, Vol. 77, 1963, pp. 559–602.

Blackwell, D. and M.A. Girshick, *Theory of Games and Statistical Decisions*, Wiley, 1954.

Bland, R.G., New finite pivoting rules for the simplex method, *Mathematics of Operations Research*, Vol. 2, 1977, pp. 103–107.

Böhm-Bawerk, E. von, *Positive Theory of Capital* (translation of 1891 German edition), Stechert, 1923.

Bomze, I., Non-cooperative two-person games in biology: A classification, *International Journal of Game Theory*, Vol. 15, 1986, pp. 31–57.

Bomze, I. and B. Pötscher, *Game Theoretical Foundations of Evolutionary Stability*, Springer, 1989.

Botvinnik, M.M., *Computers, Chess, and Long-Range Planning*, Springer, 1970.

Braithwaite, R.B., *Theory of Games as a Tool for the Moral Philosopher*, Cambridge University Press, 1955.

Brams, S., *Game Theory and Politics*, Free Press, 1975.

Brams, S., *Paradoxes in Politics*, Free Press, 1976.

Brams, S.J. and P.D. Straffin, Prisoner's dilemma and professional sports draft, *American Mathematical Monthly*, Vol. 86, 1979, pp. 80–86.

Brown, G.W., Iterative solution of games by fictitious play, in *Activity Analysis of Production and Allocation*, T.C. Koopmans (Ed.), Wiley, 1951, pp. 374–376.

Buchanan, J. M. and G. Tullock, *The Calculus of Consent*, University of Michigan Press, 1962.

Burger, E., *Introduction to the Theory of Games*, Prentice-Hall, 1963.

Camerer, C.F. *Behavioral Game Theory. Experiments in Strategic Interaction.* Princeton University Press, 2003.

Caplow, T., A theory of coalition in the triad, *American Sociological Review*, Vol. 21, 1956, pp. 489–393.

Caplow, T., Further developments of a theory of coalitions in the triad, *American Sociological Review*, Vol. 66, 1959, pp. 488–493.

Cassady, R. Jr., *Auctions and Auctioneering*, University of California Press, 1967.

Chin, H.H., T. Parsatharathy, and T.E.S. Raghavan, Structure of equilibria in n-person non-cooperative games, *International Journal of Game Theory*, Vol. 3, 1974, pp. 1–19.

Conway, J.H., *On Numbers and Games*, Academic, 1976.

Cornelius, M. and A. Parr, *What's Your Game?* Cambridge University Press, 1991.

Cowen, R. and R. Dickau, Playing games with Mathematica, *Mathematics in Education and Research*, Vol. 7, 1998, pp. 5–10.

Cowen, R. and J. Kennedy, *Discovering Mathematics with Mathematica*, Erudition Books, 2001.

Dantzig, G.B., Constructive proof of the minimax theorem, *Pacific Journal of Mathematics*, Vol. 6, 1956, pp. 25–33.

Davenport, W.C., Jamaican fishing village, *Papers in Caribbean Anthropology*, Vol. 59, Yale University Press, 1960.

Davis, M.D., *Game Theory, A Nontechnical Introduction*, Basic Books, 1970.

Dawid, H. and A. Mehlmann, Genetic learning in strategic form games, *Complexity*, Vol. 1, 1996, pp. 51–59.

Dickhaut, J. and T. Kaplan, A program for finding Nash equilibria, *The Mathematica Journal*, Vol. 1, 1991, pp. 87–93.

Dimand, M.A. and R.W. Dimand, *The History of Game Theory, Vol. I, From the Beginnings to 1945*, Routledge, 1996.

Domoryad, A.P., *Mathematical Games and Pastimes*, Pergamon, 1964.

Dresher, M., *The Mathematics of Games of Strategy: Theory and Applications*, Dover, 1981.

Driessen, T., *Cooperative Games, Solutions and Applications*, Kluwer, 1988.

Dubey, P. and L.S. Shapley, Mathematical properties of the Banzhaf power index, *Mathematics of Operations Research*, Vol. 4, 1979, pp. 99–131.

Dudeney, H., *The Canterbury Puzzles*, Dover, 1958.

Dudeney, H.E. *Amusements in Mathematics*. Dover, 1989.

Eaves, B.C., The linear complementarity problem, *Management Science*, Vol. 17, 1971, pp. 612–634.

Edgeworth, F.Y., *Mathematical Psychics*, Kegan Paul, 1881.

Epstein, R.A., *The Theory of Gambling and Statistical Logic*, Academic, 1967.

Fraenkel, A.S., How to beat your Wythoff games' opponent on three fronts, *American Mathematical Monthly*, Vol. 89, 1982, pp. 353–361.

Fraenkel, A.S., Wythoff games, continued fractions, cedar trees, and Fibonacci searches, *Theoretical Computer Science*, Vol. 29, 1984, pp. 49–73.

Fragnelli, V. et al. How to share railway infrastructure costs? in *Game Practice: Contributions From Applied Game Theory*, Kluwer, 1999.

Fréchet, M., Émile Borel, initiator of the theory of psychological games, *Econometrica*, Vol. 21, 1953, pp. 95–96.

Fréchet, M. and J. von Neumann, Commentary on the Borel Note, *Econometrica*, Vol. 21, 1953, pp. 118–127.

Friedman, L., Game-theory models in the allocation of advertising expenditures, *Operations Research*, Vol. 6, 1958, pp. 699–709.

Fudenberg, D. and D.K. Levine, *The Theory of Learning in Games*, The MIT Press, 1998.

Fudenberg, D. and J. Tirole, *Game Theory*, The MIT Press, 1991.

Gale, D., A curious Nim-type game, *American Mathematical Monthly*, Vol. 81, 1974, pp. 876–879.

Gale, D. The game of Hex and the Brouwer Fixed Point Theorem. *American Mathematical Monthly*, Vol. 86, 1979, pp. 818–827.

Gardner, M., *The Scientific American Book of Mathematical Puzzles and Diversions*, Simon & Schuster, 1959.

Gardner, M., *Sixth Book of Mathematical Games from Scientific American*, W.H. Freeman, 1971.

Gardner, M., *Mathematical Carnival*, Knopf, 1975.

Gardner, M., *Wheels, Life and Other Mathematical Amusements*, W.H. Freeman, 1983.

Gardner, M., *Penrose Tiles to Trapdoor Ciphers*, Mathematical Association of America, 1997.

Gillies, D.B., Solutions to general non-zero-sum games, in *Contributions to the Theory of Games*, Vol. 4, *Annals of Mathematics Studies* No. 40, Princeton University Press, 1959, pp. 47–85.

Greif, A., Economic history and game theory, in *Handbook of Game Theory with Economic Applications*, Vol. 3, Elsevier, 2002, pp. 1989–2024.

Grundy, P.M., Mathematics and games, *Eureka*, Vol. 2, 1939, pp. 6–8.

Guy, R.K., *Fair Game: How to Play Impartial Combinatorial Games*, COMAP Mathematical Exploration Series, 1989.

Guy, R.K. (Ed.), *Combinatorial Games*, Vol. 43, Proceedings of Symposia in Applied Mathematics. American Mathematical Society, 1991.

Guy, R., Combinatorial games, *Handbook of Combinatorics*, Vol. 2, North-Holland, 1996, pp. 2117–2162.

Halmos, P.R., The legend of John von Neumann, *American Mathematical Monthly*, Vol. 80, 1973, pp. 382–394.

Hamburger, H., *Games as Models of Social Phenomena*, W.H. Freeman, 1979.

Hammerstein, P. and R. Selten, Game theory and evolutionary biology, in *Handbook of Game Theory*, Vol. 2, R.J. Aumann and S. Hart (Eds.), North Holland, 1994, Chapter 28, pp. 929–994.

Harsanyi, J.C., Approaches to the bargaining problem before and after the theory of games: a critical discussion of Zeuthen's, Hick's, and Nash's theories, *Econometrica*, Vol. 24, 1956, pp. 144–157.

Harsanyi, J.C., On the rationality postulates underlying the theory of cooperative games, *Journal of Conflict Resolution*, Vol. 5, 1961, pp. 179–196.

Harsanyi, J.C., Rationality postulates for bargaining solutions in cooperative and non-cooperative games, *Management Science*, Vol. 9, 1962, pp. 141–153.

Harsanyi, J.C., Oddness of the number of equilibrium points: a new proof, *International Journal of Game Theory*, Vol. 2, 1973, pp. 235–250.

Harsanyi, J.C., Solutions for some bargaining games under the Harsanyi-Selten solution theory; I: Theoretical preliminaries; II: Analysis of specific games, *Mathematical Social Sciences*, Vol. 3, 1982, pp. 179–191, pp. 259–279.

Harsanyi, J.C. and R. Selten, A generalized Nash solution for two-person bargaining games with incomplete information, *Management Science*, Vol. 18, 1972, pp. 80–106.

Harsanyi, J.C. and R. Selten, *A General Theory of Equilibrium Selection in Games*, The MIT Press, 1988.

Haywood, O.G., Jr., Military decisions and game theory, *Journal of the Operations Research Society of America*, Vol. 2, 1954, pp. 365–385.

Herdon, G., N. Knoche, and C. Seidl (Eds.), *Mathematical Utility Theory: Utility Functions, Models, and Applications in the Social Sciences*, Springer, 1992.

Herstein, I.N. and J.W. Milnor, An axiomatic approach to measurable utility, *Econometrica*, Vol. 21, 1953, pp. 291–297.

Hillas, J. and E. Kohlberg, Foundations of Strategic Equilibrium, in *Handbook of Game Theory with Economic Applications*, Vol. 3, Elsevier, 2002, pp. 1597–1663.

Jones, A.J., *Game Theory: Mathematical Methods of Conflict*, Ellis Horwood, 1980.

Kalai, E. and M. Smorodinsky, Other solutions to Nash's bargaining problem, *Econometrica*, Vol. 53, 1985, pp. 513–518.

Karlin, S. *Mathematical Models and Theory in Games, Programming, and Economics*, Addison-Wesley, 1959.

Kreps, D.M., *Game Theory and Economic Modelling*, Oxford University Press, 1990.

Kuhn, H.W. (Ed.), *Classics in Game Theory*, Princeton University Press, 1997.

Kuhn, H.W., Lectures on the theory of games, *Annals of Mathematics Studies*, Vol. 37, Princeton University Press, 2003. (Based on lectures given in 1952)

Kuhn, H.W. and A.W. Tucker (Eds.), Contributions to the theory of games. I. *Annals of Mathematics Studies*, Vol. 24, Princeton University Press. 1950.

Kuhn, H.W. and S. Nasar (Eds.), *The Essential John Nash*, Princeton University Press, 2002.

Lemke, C.E. and J.T. Howson, Jr., Equilibrium points in bimatrix games, *SIAM Journal of Applied Mathematics*, Vol. 12, 1964, pp. 413–423.

Lewontin, R.C., Evolution and the theory of games, *Journal of Theoretical Biology*, Vol. 1, 1961, pp. 382–403.

Littlechild, S.C. and G.F. Thompson, Aircraft landing fees: a game theory approach, *Bell Journal of Economics*, Vol. 8, 1977, pp. 186–204.

Loomis, L.H., On a theorem of von Neumann, *Proceedings of the National Academy of Sciences, USA*, Vol. 32, 1946, pp. 213–215.

Lucas, W.F., A game with no solution, *Bulletin of the American Mathematical Society*, Vol. 74, 1968, pp. 237–239.

Lucas, W.F., The proof that a game may not have a solution, *Transactions of the American Mathematical Society*, Vol. 137, 1969, pp. 219–229.

Lucas, W.F., An overview of the mathematical theory of games, *Management Science*, Vol. 18, 1972, pp. 3–19.

Lucas, W., Multiperson cooperative games, in *Game Theory and its Applications*, American Mathematical Society, 1981, pp. 1–17.

Lucas, W., Applications of cooperative games to equitable allocation, in *Game Theory and Its Applications*, American Mathematical Society, 1981, pp. 19–36.

Lucas, W., Measuring power in weighted voting systems, in *Political and Related Models*, S. Brams, W. Lucas, and P. Straffin (Eds.), Springer, 1983, Chapter 9.

Lucas, W.F. and M. Rabie, Games with no solutions and empty core, *Mathematics of Operations Research*, Vol. 7, 1982, pp. 491–500.

Lucas, W.F. and R.M. Thrall, N-person games in partition function form, *Naval Logistics Research Quarterly*, Vol. 10 , 1963, pp. 281–298.

Luce, R.D. and H. Raiffa, *Games and Decisions*, John Wiley, 1957.

Luce, R.D. and A. Rogow, A game-theoretic analysis of congressional power distribution for a stable two-party system, *Behavioral Science*, Vol. 1, 1956, pp. 83–95.

Mangasarian, O.L., Equilibrium points in bimatrix games, *Journal of the Society for Industrial and Applied Mathematics* Vol. 12, 1964, pp. 778–780.

Mann, I. and L.S. Shapley, The *a priori* voting strength of the electoral college, *American Political Science Review*, Vol. 72, 1962, pp. 70–79.

Maynard, S.J., *Evolution and the Theory of Games*, Cambridge University Press, 1982.

Maynard, S.J. and G. Price, The logic of animal conflict, *Nature*, Vol. 246, 1973, pp. 15–18.

McDonald, J., *Strategy in Poker, Business, and War*, W.W. Norton, 1950.

McKelvey, R.D. and A. McLennan, Computation of equilibria in finite games, *Handbook of Computational Economics*, Vol. 1, 1996, pp. 87–142.

McKinsey, J.C.C., *Introduction to the Theory of Games*, McGraw-Hill, 1952. (Dover Reprint, 2003)

Mead, E., A. Rosa, and C. Huang, The game of SIM: A winning strategy for the second player. *Mathematics Magazine*, Vol. 47, 1974, pp. 243–247.

Mehlmann, A., *The Game's Afoot! Game Theory in Myth and Paradox*, American Mathematical Society, 2000.

Mesterton-Gibbons, M., *An Introduction to Game-Theoretic Modelling*, Addison-Wesley, 1992.

Milnor, J.A Nobel Prize for John Nash, *The Mathematical Intelligencer*, Vol. 17, 1995, pp. 14–15.

Moore, E.H., A generalization of the game called Nim, *Annals of Mathematics*, Vol. 11, 1910, pp. 93–94.

Morris, P., *Introduction to Game Theory*, Springer, 1994.

Moulin, H., *Game Theory for the Social Sciences*, New York University Press, 1986.

Myerson, R., Refinements of the Nash equilibrium concept, *International Journal of Game Theory*, Vol. 7, 1978, pp. 73–80.

Myerson, R., *Game Theory: Analysis of Conflict*, Harvard University Press, 1991.

Nachbar, J.H., Evolution in the finitely repeated Prisoner's dilemma, *Journal of Economic Behavior and Organization*, Vol. 19, 1992, pp. 307–326.

Nasar, S. *A Beautiful Mind*. Simon & Schuster, 1998.

Nash, J.F. Jr., Equilibrium points in *n*-person games, *Proceedings of the National Academy of Sciences*, Vol. 36, 1950a, pp. 48–49. (Reprinted in Kuhn [1997])

Nash, J.F. Jr., The bargaining problem, *Econometrica*, Vol. 18, 1950b, pp. 155–162. (Reprinted in Kuhn [1997])

Nash, J.F. Jr., Non-cooperative games, *Annals of Mathematics*, Vol. 54, 1951, pp. 286–295. (Reprinted in Kuhn [1997])

Nash, J.F. Jr., Two-person cooperative games, *Econometrica*, Vol. 21, 1953, pp. 128–140.

Nowakowski, R. (Ed.), *Games of No Chance*, Cambridge University Press, 1996.

Nowakowski, R. (Ed.), *More Games of No Chance*, Cambridge University Press, 2003.

O'Beirne, T.H., *Puzzles and Paradoxes*, Dover, 1984. (Oxford University Press, 1965)

Osborne, M.J. and A. Rubinstein, *A Course in Game Theory*, The MIT Press, 1994.

Owen, G., *Game Theory*, W.B. Saunders, 1968; (2nd ed.) Academic Press, 1982.

Owen, G., Evaluation of a presidential election game, *American Political Science Review*, Vol. 69, 1975, pp. 947–953.

Parthasarathy, T. and T.E.S. Raghavan, *Some Topics in Two-Person Games*, Elsevier, 1971.

Pierce, J.R., *Symbols, Signals, and Noise*, Harper, 1961.

Peleg, B., *Game Theoretic Analysis of Voting in Committees*, Cambridge University Press, 1984.

Rapoport, A., *Two-Person Game Theory: The Essential Ideas*, University of Michigan Press, 1966.

Rapoport, A., *N-Person Game Theory: Concepts and Applications*, University of Michigan Press, 1970.

Rasmusen, E., *Games and Information* (2nd ed.), Basil Blackwell, 1994.

Riker, W.H., *The Theory of Political Coalitions*, Yale University Press, 1962.

Robinson, J., An iterative method of solving a game, *Annals of Mathematics*, Vol. 54, 1951, pp. 296–301.

Rosenmüller, J., On a generalization of the Lemke–Howson algorithm to non-cooperative N-person games, *SIAM Journal of Applied Mathematics*, Vol. 21, 1971, pp. 73–79.

Rosenthal, R., Games of perfect information, predatory pricing, and the chain-store paradox, *Journal of Economic Theory*, Vol. 25, 1981, pp. 92–100.

Ross, D., *What People Want: The Concept of Utility from Bentham to Game Theory*, University of Capetown Press, 1999.

Roth, A.E., The evolution of the labor market for medical interns and residents: a case study in game theory, *Journal of Political Economy*, Vol. 92, 1984, pp. 991–1016.

Roth, A.E., *Game-Theoretic Models of Bargaining*, Cambridge University Press, 1985.

Roth, A. (Ed.), *The Shapley Value: Essays in Honor of Lloyd S. Shapley*, Cambridge University Press, 1988.

Samuelson, L., *Evolutionary Games and Equilibrium Selection*, The MIT Press, 1997.

Scarf, H.E., The core of an *n*-person game, *Econometrica*, 1967, pp. 50–69.

Scarf, H.E., *The Computation of Economic Equilibria*, Yale University Press, 1973.

Schelling, T.C., *The Strategy of Conflict*, Harvard University Press, 1960.

Schotter, A. and G. Schwödiauer, Economics and the theory of games, *Journal of Economic Literature*, Vol. 18, 1980, pp. 479–527.

Schuh, F., *The Master Book of Mathematical Recreations*, Dover, 1968

Schwartz, B. S. (Ed.), *Mathematical Solitaires and Games*, Baywood, 1980.

Scodel A. et al., Some descriptive aspects of two-person non-zero-sum games, *The Journal of Conflict Resolution*, Vol. 3, 1959, pp. 114–119.

Scodel, A. et al., Some descriptive aspects of two-person non-zero-sum games, *The Journal of Conflict Resolution*, Part II, Vol. 4, 1960, pp. 193–197.

Shader, L. Another strategy for SIM, *Mathematics Magazine*, Vol. 51, 1978, pp. 60–64.

Shapley, L.S., A value for *n*-person games, *Contributions to the Theory of Games II*, Princeton University Press, 1953, pp. 307–317 (Reprinted in Kuhn [1997])

Shapley, L., Valuation of games, in *Game Theory and Its Applications*, American Mathematical Society, 1981, pp. 55–67.

Shapley, L., Measurement of power in political systems, in *Game Theory and Its Applications*, American Mathematical Society, 1981, pp. 69–81.

Shapley, L.S. and M. Shubik, A method for evaluating the distribution of power in a committee system, *The American Political Science Review*, Vol. 48, 1954, pp. 787–792.

Shubik, M., The role of game theory in economics, *Kyklos*, Vol. 6, 1953, pp. 21–34.

Shubik, M. (Ed.), *Readings in Game Theory and Political Behavior*, Doubleday, 1954.

Shubik, M., *Strategy and Market Structure*, Wiley, 1959.

Shubik, M., *Games for Society, Business, and War*, Elsevier, 1975.

Shubik, M., *Game Theory in the Social Sciences: Concepts and Solutions*, The MIT Press, 1982.

Silverman, D.L., *Your Move*, Kaye and Ward, 1971.

Sorin, S., *A First Course on Zero-Sum Repeated Games*, Springer, 2002.

Sprague, R.P., Ueber mathematische Kampfspiele, *Tohoku Mathematical Journal*, Vol. 41, 1935–36, pp. 438–444.

Sprague, R., *Recreation in Mathematics*, Dover, 1963.

Straffin, P., *Topics in the Theory of Voting*, Birkhauser, 1980.

Straffin, P., Power indices in politics, in *Political and Related Models*, S. Brams, W. Lucas, and P. Straffin (Eds.), Springer, 1983, Chapter 11.

Straffin, P. D., *Game Theory and Strategy*, Mathematical Association of America, 1993.

Straffin, P.D., Power and stability in politics, in *Handbook of Game Theory with Economic Applications*, Vol. 2, North-Holland, 1994, pp. 1127–1152.

Sultan, A., *Linear Programming*, Academic Press, 1993.

Thomas, L.C., *Games, Theory and Applications*, Ellis Horwood, 1984.

Tan, T. and S. Werlang, The Bayesian foundations of solution concepts of games, *Journal of Economic Theory*, Vol. 45, 1988, pp. 370–391.

Tirole, J., *The Theory of Industrial Organization*, The MIT Press, 1988.

Vajda, S., *Mathematical Games and How to Play Them*, Ellis Horwood, 1992.

Van Damme, E., *Stability and Perfection of Nash Equilibria*, Springer, 1983.

Van Damme, E., Strategic Equilibrium, in *Handbook of Game Theory with Economic Applications*, Vol. 3, Elsevier, 2002, pp. 1521–1596.

Vega-Redondo, F., *Evolution, Games, and Economic Behavior,* Oxford University Press, 1996.

Vickrey, W., Auctions and bidding games, in *Recent Advances in Game Theory,* Princeton University Press, 1962, pp. 15–27.

Ville, J., Sur la théorie générale des jeux où intervient l'habilité des joueurs, in *Applications des Jeux de Hasard,* Vol. 4, E. Borel et al. (1925–1939), 1938, Fasc. 2, pp. 105–113.

Von Neumann, J., Zur Theorie der Gesellschaftsspiele, *Math. Annalen,* Vol. 100, 1928, pp. 295–320. (English translation by S. Bargmann, On the theory of games of strategy, in *Contributions to the Theory of Games,* Vol. 4, A.A. Tucker and R.D. Luce (Eds.), *Annals of Mathematical Studies,* Vol. 40, Princeton University Press, 1959.)

Von Neumann, J. and O. Morgenstern, *Theory of Games and Economic Behavior,* Princeton University Press, 1944 (60th Anniversary Edition, 2004)

Von Stengel, B., Computing equilibria for two-person games, *Handbook of Game Theory with Economic Applications,* Vol. 3, Elsevier, 2002, pp. 1723–1759.

Weber, R., Games in coalitional form, in *Handbook of Game Theory,* Vol. 2, North-Holland, 1994, pp. 1285–1304.

Weibull, J.W., *Evolutionary Game Theory,* The MIT Press, 1995.

Weyl, H., Elementary proof of a Minimax Theorem due to von Neumann, in *Contribution to the Theory of Games,* Vol. 1, Kuhn, H.W. and A.W. Tucker (Eds.), 1950, pp. 19–25.

Williams, J.D., *The Compleat Strategyst: Being a Primer on the Theory of Games of Strategy,* Dover, 1986.

Wilson, R., Computing equilibria of n-person games, *SIAM Journal of Applied Mathematics,* Vol. 21, 1971, pp. 80–87.

Winkels, H.-M., An algorithm to determine all equilibrium points of a bimatrix game, in *Game Theory and Related Topics,* O. Moeschler and D. Pallaschke (Eds.), North-Holland, 1979, pp. 137–148.

Winter, E., The Shapley value, in *Handbook of Game Theory with Economic Applications,* Vol. 3, Elsevier, 2002, pp. 2025–2054.

Wythoff, W.A., A modification of the game of Nim, *Nieuw Archief voor Wiskunde,* Vol. 7, 1907, pp. 199–202.

Yaglom, A.M. and I.M. Yaglom, *Challenging Mathematical Problems with Elementary Solutions,* Dover, 1987. (Original Russian edition, Moscow, 1954)

Young, H.P., Cost allocation, in *Handbook of Game Theory,* Vol. 2, North-Holland, 1994, pp. 1193–1235.

Zeeman, E.C., Population dynamics from game theory, in *Global Theory of Dynamic Systems,* Springer, 1979, pp. 471–497.

Zermelo, E., Ueber eine Anwendung der Mengenlehre auf die Theorie des Schachspiels, *Proceedings of the Fifth International Congress of Mathematicians,* Vol. 2, Cambridge, 1912, pp. 501–504.

Zeuthen, F., *Problems of Monopoly and Economic Welfare,* Routledge & Kegan Paul, 1930.

Zwicker, W.S., Playing games with games: the Hypergame paradox, *American Mathematical Monthly,* 1987, pp. 507–514.

Index